Supporting Air and Space Expeditionary Forces

Expanded Operational Architecture for Combat Support Execution Planning and Control

Patrick Mills, Ken Evers, Donna Kinlin, Robert S. Tr[illegible]

Prepared for the United States Air Force
Approved for public release; distribution unlimited

RAND PROJECT AIR FORCE

The research described in this report was sponsored by the United States Air Force under Contract F49642-01-C-0003. Further information may be obtained from the Strategic Planning Division, Directorate of Plans, Hq USAF.

Library of Congress Cataloging-in-Publication Data

Supporting air and space expeditionary forces : expanded operational architecture for combat support execution planning and control / Patrick Mills ... [et al.].
p. cm.
"MG-316."
Includes bibliographical references.
ISBN 0-8330-3838-9 (pbk. : alk. paper)
1. United States. Air Force—Supplies and stores. 2. Deployment (Strategy) 3. Command and control systems—United States. 4. Air power—United States. 5. Air warfare. I. Mills, Patrick, 1975–

UG1103.S89 2005
355.4'1415'0973—dc22

2005023053

Published 2006 by the RAND Corporation
1776 Main Street, P.O. Box 2138, Santa Monica, CA 90407-2138
1200 South Hayes Street, Arlington, VA 22202-5050
4570 Fifth Avenue, Suite 600, Pittsburgh, PA 15213
RAND URL: http://www.rand.org/
To order RAND documents or to obtain additional information, contact
Distribution Services: Telephone: (310) 451-7002;
Fax: (310) 451-6915; Email: order@rand.org

Preface

This report expands and provides more detail on several organizational nodes in our earlier work that outlined concepts for an operational architecture for guiding the development of Air Force combat support (CS) execution planning and control needed to enable rapid deployment and employment of the Air and Space Expeditionary Force (AEF). These CS execution planning and control processes are sometimes referred to as CS command and control (CSC2) processes. We will use CSC2 to describe these processes in this report.

This work was conducted by the Resource Management Program of RAND Project AIR FORCE and was sponsored jointly by the USAF Deputy Chief of Staff of Installations and Logistics (AF/IL) and the Commander of Air Force Materiel Command (AFMC/CC). It is one of a series of analyses addressing how best to support Expeditionary Air and Space Forces. Other reports in this series include:

- *Supporting Expeditionary Aerospace Forces: An Integrated Strategic Agile Combat Support Planning Framework*, Robert S. Tripp et al. (MR-1056-AF, 1999)
- *Supporting Expeditionary Aerospace Forces: New Agile Combat Support Postures*, Lionel Galway et al. (MR-1075-AF, 2000)
- *Supporting Expeditionary Aerospace Forces: An Analysis of F-15 Avionics Options*, Eric Peltz et al. (MR-1174-AF, 2000)

- *Supporting Expeditionary Aerospace Forces: A Concept for Evolving the Agile Combat Support/Mobility System of the Future*, Robert S. Tripp et al. (MR-1179-AF, 2000)
- *Supporting Expeditionary Aerospace Forces: Expanded Analysis of LANTIRN Options*, Amatzia Feinberg et al. (MR-1225-AF, 2001)
- *Supporting Expeditionary Aerospace Forces: Lessons from the Air War Over Serbia*, Amatzia Feinberg et al. (2002, government publication; not releasable to the general public)
- *Supporting Expeditionary Aerospace Forces: Alternatives for Jet Engine Intermediate Maintenance*, Mahyar A. Amouzegar, Lionel R. Galway, and Amanda Geller (MR-1431-AF, 2002)
- *Supporting Expeditionary Aerospace Forces: An Operational Architecture for Combat Support Execution Planning and Control*, James Leftwich et al. (MR-1536-AF, 2002)
- *Supporting Expeditionary Aerospace Forces: Lessons from Operation Enduring Freedom*, Robert S. Tripp et al. (MR-1819-AF, 2004).
- *Supporting Air and Space Expeditionary Forces: Analysis of Maintenance Forward Support Location Operations*, Amanda Geller et al. (MG-151-AF, 2004)
- *Supporting Air and Space Expeditionary Forces: A Methodology for Determining Air Force Deployment Requirements*, Don Snyder and Patrick Mills (MG-176-AF, 2004)
- *Supporting Air and Space Expeditionary Forces: Analysis of Combat Support Basing Options*, Mahyar A. Amouzegar et al. (MG-261-AF, 2004)
- *Supporting Air and Space Expeditionary Forces: Lessons from Operation Iraqi Freedom*, Kristin F. Lynch et al. (MG-193-AF, 2005).

This report should be of interest to commanders, logisticians, and planners in AFMC and AF/IL, as well as Commanders of Air Force Forces (COMAFFORs) and their A-Staffs.

RAND Project AIR FORCE

RAND Project AIR FORCE (PAF), a division of the RAND Corporation, is the U.S. Air Force's federally funded research and development center for studies and analyses. PAF provides the Air Force with independent analyses of policy alternatives affecting the development, employment, combat readiness, and support of current and future aerospace forces. Research is conducted in four programs: Aerospace Force Development; Manpower, Personnel, and Training; Resource Management; and Strategy and Doctrine.

Additional information about PAF is available on our Web site at http://www.rand.org/paf.

Contents

Figures

Tables

Summary

Introduction and Motivation

During the past decade, the U.S. military has supported continuous deployments of forces around the world, often on very short notice and for prolonged duration, to meet the needs of a wide range of peacekeeping and humanitarian missions or major contingency operations. The pattern of varied and fast-breaking regional crises appears to be the model for the foreseeable future and has prompted the United States to reassess how it prepares, maintains, and employs its military forces.[1] In response to this operating environment, the Air Force has reorganized into an Air and Space Expeditionary Force (AEF).

The AEF concept divides the Air Force into ten relatively equal groups (i.e., AEFs) of people and equipment. In any given 90-day period, two AEFs (or one AEF pair) are vulnerable to deployment to fulfill steady-state Air Force deployment requirements.[2] The aim of this concept is to replace a permanent forward presence with forces that are primarily stationed in the continental United States (CONUS) and can be tailored rapidly, deployed quickly, employed immediately, and sustained indefinitely.

---

[1] Donald Rumsfeld, *Defense Strategy Review*, June 21, 2001; Donald Rumsfeld, *Guidance and Terms of Reference for the 2001 Quadrennial Defense Review*, June 22, 2001.

[2] Some assets are not easily divided into ten AEFs and are therefore managed separately, as "enablers" (e.g., AWACS [Airborne Warning and Control System], strategic mobility). These assets are on call at all times.

These AEF global force projection goals present significant challenges to the current combat support (CS) system.[3] CS is the collection of people, equipment, and processes that create, protect, and sustain air and space forces across the full range of military operations.[4] In addition to the importance of CS, command and control (C2) has been identified as a key component of the AEF Agile Combat Support (ACS) system that needs further development.[5] Joint doctrine defines C2 as the exercise of authority and direction by a properly designated commander over assigned and attached forces in the accomplishment of the mission.[6] CSC2,[7] thus, is the exercise of authority and direction (planning, directing, coordinating, and controlling[8]) over CS forces and resources to meet operational objectives.

To date, operational planning has not sufficiently incorporated CSC2. It is challenging to do so for several reasons. CS planners usually do not have up-to-date and reliable CS resource information in a format that can be easily broken down for use by operators. Also,

---

3 Throughout this report, we use "system" in the general sense—a combination of facts, principles, methods, processes, and the like. We use the expression information system to refer specifically to a product designed to manage data.

4 ACS concept of operations (CONOPS), January 21, 2005. The CONOPS document includes many functions in CS, such as civil engineering, communications and information, logistics readiness, maintenance, munitions, and security forces.

5 Research at the RAND Corporation has focused on defining the vision and evaluating options for an ACS system that can meet AEF operational goals. See Lionel A. Galway, Robert S. Tripp, Timothy L. Ramey, and John G. Drew, *Supporting Expeditionary Aerospace Forces: New Agile Combat Support Postures*, Santa Monica, Calif.: RAND Corporation, MR-1075-AF, 2000. Additional research has identified the importance of CSC2 within the AEF ACS system. See Robert S. Tripp, Lionel A. Galway, Paul S. Killingsworth, Eric Peltz, Timothy L. Ramey, and John G. Drew, *Supporting Expeditionary Aerospace Forces: An Integrated Strategic Agile Combat Support Planning Framework*, Santa Monica, Calif.: RAND Corporation, MR-1056-AF, 1999.

6 Joint Chiefs of Staff, Joint Publication 1-02, *Department of Defense Dictionary of Military and Associated Terms*, Department of Defense, April 12, 2001.

7 This report deals with the processes associated with CS execution planning and control. Often these processes have been referred to as the set of combat support command and control (CSC2) processes. We will use CSC2 to describe these processes in this report.

8 Department of the Air Force, Air Force Doctrine Document 2-8, *Command and Control*, February 16, 2001.

most logisticians are not trained in and do not participate in air campaign planning. Finally, operators are often unwilling to commit early on to plans (to the degree that they would put them in writing and pass them on to CS planners).

This work expands on the work of Leftwich et al.,[9] which presented initial concepts for guiding the development of a CSC2 operational architecture[10] for the AEF. When that work was started, the Air Force simply had no operational architecture for CSC2. Leftwich addressed the problem of CS not being integrated into operational planning, focusing mostly on the Commander of Air Force Forces (COMAFFOR) and Joint Forces Air Component Commander (JFACC) levels during strategic planning and contingency planning and execution. For example, during crisis action planning, Air Force operators had limited access to CS information to influence their decisions.[11] The Air Force began to implement Leftwich's recommendations but asked for further work. The current work introduces new concepts for Air Force involvement in the planning, programming, budgeting, and execution processes and provides further detail on CS contingency planning and execution processes associated with

---

[9] Research at RAND defined an initial concept for a CS execution planning and control architecture. See James Leftwich, Robert Tripp, Amanda Geller, Patrick Mills, Tom LaTourrette, and C. Robert Roll, Jr., *Supporting Expeditionary Aerospace Forces: An Operational Architecture for Combat Support Execution Planning and Control,* Santa Monica, Calif.: RAND Corporation, MR-1536-AF, 2002.

[10] An operational architecture, within the Department of Defense (DoD), is a description of tasks, operational elements, and information flows required to accomplish or support a DoD function or military operation. It describes the operational elements, assigned tasks and activities, and information flows required to support the warfighter. It defines the types of information exchanged, the frequency of exchange, which tasks and activities are supported by the information exchanges, and the nature of information exchanges in sufficient detail to ascertain specific interoperability requirements (Department of Defense, *C4ISR Framework Document Version 2.0,* December 18, 1997). The Leftwich report and this report are not by themselves, nor do they contain, operational architectures per se. The results of these analyses are concepts that may guide the Air Force in developing and refining its CSC2 operational architecture. For simplicity's sake, we refer to our results as an "operational architecture," although they are in fact concepts for such architecture.

[11] During Operation Allied Force, the single CS person responsible for interaction in the operational planning group did not have a full depth of CS experience, information system links, or decision support tools to help facilitate interaction.

specific organizational nodes described in the earlier report. While Leftwich described some of the CSC2 information produced and passed between organizational nodes, this work goes further in depth and breadth, adding detail on those information flows and the decisions they drive.[12] We intend this study to guide the refinement of the Air Force's CSC2 operational architecture, specifically at the Headquarters Air Force Materiel Command and COMAFFOR A-Staff levels.[13]

We recognize that coalition support has been a key factor in recent U.S. military operations and that coordinating and integrating the CS of coalition partners has been a significant challenge. However, that issue was outside the scope of this report—which focuses on internal Air Force issues—and is not treated here.

Throughout this report, we refer to four different operational architectures: the AS-IS, the TO-BE, the Evolving AS-IS, and the Expanded TO-BE. Leftwich et al. took as their starting point the existing operational architecture, calling it the AS-IS. The results of their research were assembled into what they called the future, or TO-BE. Because the actual operational architecture they observed has evolved since the original work—due to Air Force–initiated changes and implementations of some of Leftwich's concepts—we refer to the current architecture that we took as the starting point for our analysis as the Evolving AS-IS. We analyzed the Evolving AS-IS architecture and built on some of Leftwich et al.'s more general architectural concepts. We refer to the assembly of our results as the Expanded TO-BE.

---

[12] The CD-ROM enclosed with this document contains a library of dozens of proposed information products for several different organizational nodes.

[13] Rather than view the results of this study as a CSC2 operational architecture, which would promote the concept of a stovepiped, nonintegrated architecture, we address CS execution planning and control processes in the context of the larger Air Force C2 architecture.

Analytic Approach

Our study builds on a previous one that developed initial concepts for a future (TO-BE) operational architecture. We analyzed the Evolving AS-IS CSC2 architecture, identified changes needed in this architecture to realize AEF operational goals and correct deficiencies identified during recent contingencies, and expanded on the previous concepts in the TO-BE architecture. The concepts in this report incorporate evolving practices; information from interviews with Air Force personnel; lessons from the operations Allied Force, Enduring Freedom, and Iraqi Freedom; and results of the authors' analysis of the current CSC2.

CSC2 Recommendations to Meet the TO-BE Architecture

The Air Force has already initiated changes aimed at implementing doctrine and policy changes according to the TO-BE operational architecture, and plans are in place to continue to close the gaps. Our analysis of the Air Force's CS execution planning and control process revealed remaining shortfalls in the Evolving AS-IS architecture, including the following:

- Operational parameters are not consistently communicated to CS planners early in crisis action planning. (See pages 40–42.)
- Capability assessments are often conducted on an ad hoc basis. (See page 42.)
- Oversight for personnel and equipment resources is spread across multiple organizations. (See page 46.)
- The Spares Commodity Control Point (CCP) lacks closed-loop planning and execution processes and mechanisms. (See pages 46–50.)
- The Combat Support Center (CSC) has limited analytic capability. (See pages 50–51.)

- The deployment planning system lacks the ability to plan or re-plan and to rapidly explore multiple deployment options. (See pages 53–57.)

We propose an Expanded TO-BE CSC2 architecture that would enable the Air Force to meet its AEF operational goals.

Finally, this report offers the following recommendations to help the Air Force CS community move from the current architecture to the future concept we describe:

- An operational parameters template and capability assessment concepts should be codified in Air Force CS doctrine and policy. Creating a framework, reinforced in doctrine, to delineate specifically what information operators provide, in what format, to CS planners during crisis action planning is necessary to improve the coordination, timeliness, and accuracy of CS planning. The content and format of capability assessments should be codified in doctrine and policy. (See pages 40–42.)
- Personnel and equipment oversight should be brought under one organization to simplify accountability and make deployment planning more efficient. (See page 46.)
- Analytic capabilities should be enhanced in the CSC. (See pages 50–51.)
- Standing CS organizations should be enhanced to further enable execution planning and control. A closed-loop feedback process incorporating depot maintenance and the program objective memorandum process should be included in spares CCP operations. Personnel and equipment oversight should be brought under one organization to simplify accountability and make deployment planning more efficient. Analytic capability should be added to the CSC. (See pages 42–51.)
- Trained operators are needed to create, and CS planners to effectively use, operational parameters templates. The concept and usage of the Operational Parameters Template delineated in doctrine should be reinforced by training operators and CS planners in its design and use. (See pages 51–53.)

- Appropriate information system and decision support tools should be fielded to meet Expanded TO-BE architectural requirements. This will increase access to capabilities assessments, better connect spares planning and execution, and improve the deployment planning process. (See pages 53–57.)

Conclusion

The strategic and operational environment and the AEF concept that addresses it present significant challenges to the current CS structure. Correcting remaining deficiencies in CSC2 as identified in this report is integral to the continued success of this effort.

Acknowledgments

This project was sponsored by both AF/IL and AFMC/CC; we thank our sponsors, Lieutenant General Michael Zettler (Ret.) and General Lester L. Lyles (Ret.), for their support of this work.

At the Air Staff, we are especially grateful to Grover Dunn (AF/ILI), who took personal interest in the project and participated in updating briefings. We also thank Lieutenant General Terry Gabreski (AFMC/CV); Brigadier General David Gillett (AF/ILM); Major General Art Morrill (AFMC/LG); Brigadier General Polly Peyer (PACAF/LG); Colonel Russ Grunch (Ret.) (PACAF/LG-ALOC); Mr. Curt Newmann (AFMC/XP); and their staffs for their support and critique of this work. Major General Morrill was instrumental in setting up an assessment framework that is being used in ongoing exercises for evaluating the extent to which the concepts in this architecture have been implemented. We thank Wing Commander David Orr and Major Maria Garcia for arranging our visits to AFMC and USAFE, respectively.

We particularly thank our colleague Charles Robert Roll, Jr., for his guidance and leadership in the project.

Finally, we wish to thank our Air Force action officer, Colonel Connie Morrow (Ret.) (AF/ILGX). We also thank Mr. Dick Olson (AF/ILGX) for his support and exchange of ideas. At RAND, Amanda Geller made key contributions to the research reported here. Also at RAND, we give thanks to Lawrence Hanser and Myron Hura for reviewing this document and supplying helpful comments; June Kobashigawa for helping to prepare this document; and Sandra

Wade-Grusky, Dan Agostino, and Neal Sofge for preparing the HTML flowcharts and product library.

We, of course, assume responsibility for any errors or omissions.

Abbreviations

A-3	Director of Operations for Air Force Forces
A-4	Director of Logistics for Air Force Forces
A-5	Director of Plans and Programs for Air Force Forces
A-6	Director of Communications and Information for Air Force Forces
A-7	Director of Installations and Mission Support for Air Force Forces
ACC	Air Combat Command
ACES	Agile Combat Execution Support
ACS	Agile Combat Support
AEF	Air and Space Expeditionary Force
AEFC	Air Expeditionary Force Center
AF/IL	Office of the Deputy Chief of Staff of the Air Force for Installations and Logistics
AFDD	Air Force Doctrine Document
AFMC	Air Force Materiel Command
AFMC/LSO	Air Force Materiel Command Logistics Support Office
ALC	air logistics center
AOC	air operations center
AOR	area of responsibility

APPG	Air Force Planning and Programming Guidance
APS	Advanced Planning and Scheduling
AWACS	Airborne Warning and Control System
BEAR	basic expeditionary airfield resources
BES	Budget Estimate Submission
C2	command and control
CCP	Commodity Control Point
CENTAF	Central Air Forces
CENTCOM	Central Command
CIRF	Centralized Intermediate Repair Facility
COA	course of action
COMAFFOR	Commander of Air Force Forces
CONOPS	concept of operations
CONUS	continental United States
CS	combat support
CSC	Combat Support Center
CSC2	combat support command and control
CSL	CONUS support location
CWT	customer wait time
DLR	depot-level reparable
DoD	Department of Defense
EPM	EXPRESS Planning Module
EXPRESS	Execution and Prioritization of Repair Support System
FMSE	Fuels Mobility Support Equipment
FOL	forward operating location
FSL	forward support location
FYDP	Future Years Defense Program
HQ-AF	Headquarters Air Force

HTML	Hypertext Markup Language
JDAM	Joint Direct Attack Munition
JFACC	Joint Forces Air Component Commander
ILGX	Logistics Readiness Directorate, Office of the Deputy Chief of Staff for Installations and Logistics
LG	Logistics Group
LG-ALOC	Logistics Group–Air Logistics Operations Center
LSA	logistics supportability analysis
LSC	logistics support center
MAJCOM	major command
MDS	mission design series
MOE	measure of effectiveness
MSG	Mission Support Group
NAF	numbered air force
OAF	Operation Allied Force
OEF	Operation Enduring Freedom
OIF	Operation Iraqi Freedom
ONW	Operation Northern Watch
OPT	Operational Parameters Template
OSC	Operations Support Center
OSD	Office of the Secretary of Defense
OSW	Operation Southern Watch
PACAF	Pacific Air Forces
POL	petroleum, oil, and lubricants
POM	program objective memorandum
POSC	Pacific Operations Support Center
PPBE	Planning, Programming, Budgeting, and Execution
RSS	regional supply squadron

SOS/R	source of supply or repair
SRRB	Spares Requirements Review Board
TDS	Theater Distribution System
TPFDD	Time-Phased Force and Deployment Data
USAFE	United States Air Forces, Europe
USTRANSCOM	United States Transportation Command
UTASC	USAFE Theater Air Support Center
UTC	unit type code
WRM	war reserve materiel
WSA	weapon system availability
WS-SCM	weapon system supply chain manager

CHAPTER ONE

Introduction

During the past decade, the U.S. military has supported continuous deployments of forces around the world, often on very short notice and for prolonged duration, to meet the needs of a wide range of peacekeeping and humanitarian missions or major contingency operations. The pattern of varied and fast-breaking regional crises appears to be the model for the foreseeable future and has prompted the United States to reassess how it prepares, maintains, and employs its military forces.[1] In response to this operating environment, the Air Force has reorganized into an Air and Space Expeditionary Force (AEF).

The AEF concept divides the Air Force into ten relatively equal groups (i.e., AEFs) of people and equipment. In any given 90-day period, two AEFs (or one AEF pair) are vulnerable to deployment to fulfill steady-state Air Force deployment requirements.[2] The aim of this concept is to replace a permanent forward presence with forces that are primarily stationed in the continental United States (CONUS) and can be tailored rapidly, deployed quickly, employed immediately, and sustained indefinitely.

[1] Donald Rumsfeld, *Defense Strategy Review*, June 21, 2001a; Donald Rumsfeld, *Guidance and Terms of Reference for the 2001 Quadrennial Defense Review*, June 22, 2001b.

[2] Some assets are not easily divided into ten AEFs and are therefore managed separately, as "enablers" (e.g., AWACS [Airborne Warning and Control System], strategic mobility). These assets are on call at all times.

These AEF global force projection goals present significant challenges to the current combat support (CS) system.[3] CS is the collection of people, equipment, and processes that create, protect, and sustain air and space forces across the full range of military operations.[4] It spans many functional areas, including civil engineering, communications and information, logistics readiness, maintenance, munitions, and security forces.[5]

In addition to challenges to the CS system, the importance of command and control (C2) has been identified as a key component of the AEF Agile Combat Support (ACS) system that needs further development.[6] Joint doctrine defines C2 as the exercise of authority and direction by a properly designated commander over assigned and attached forces in the accomplishment of the mission.[7] Air Force doctrine includes in it the functions of planning, directing, coordinating, and controlling forces and resources.[8] In this report, we

---

[3] Throughout this report, we use the word system in the general sense—a combination of facts, principles, methods, processes, and the like. We use the expression information system to refer specifically to a product designed to manage data.

[4] Definition of ACS concept of operations (CONOPS), January 21, 2005.

[5] The ACS CONOPS (January 21, 2005) also lists Acquisitions, Airfield Operations, Chaplain, Contracting, Financial Management and Comptroller, Health Services, Historian, Judge Advocate General, Manpower, Office of Special Investigations, Personnel, Postal Service, Public Affairs, Safety, Science and Technology, Services, Test and Evaluation.

[6] Research at RAND has focused on defining the vision and evaluating options for an ACS system that can meet AEF operational goals. See Lionel A. Galway, Robert S. Tripp, Timothy L. Ramey, and John G. Drew, *Supporting Expeditionary Aerospace Forces: New Agile Combat Support Postures*, Santa Monica, Calif.: RAND Corporation, MR-1075-AF, 2000. Additional research has identified the importance of CSC2 within the AEF ACS system. See Robert S. Tripp, Lionel A. Galway, Paul S. Killingsworth, Eric Peltz, Timothy L. Ramey, and John G. Drew, *Supporting Expeditionary Aerospace Forces: An Integrated Strategic Agile Combat Support Planning Framework*, Santa Monica, Calif.: RAND Corporation, MR-1056-AF, 1999.

[7] Joint Chiefs of Staff, Joint Publication 1-02, *Department of Defense Dictionary of Military and Associated Terms*, Department of Defense, April 12, 2001.

[8] Definition of C2 from Department of Air Force, Air Force Doctrine Document 2-8, *Command and Control*, February 16, 2001. For more detail on how these four activities can be applied to CS, see Chapter Five of James Leftwich, Robert Tripp, Amanda Geller, Patrick Mills, Tom LaTourrette, and C. Robert Roll, Jr., *Supporting Expeditionary Aerospace Forces:*

expand on joint and Air Force definitions of C2, typically applied to battlespace management, and address the functions of planning, directing, coordinating, and controlling *CS resources* to meet operational objectives—in other words, CSC2.[9] In a narrow sense, these standard definitions, because they deal with battlespace management, include C2 functions with respect to the operational and tactical levels of warfare. We take a wider view and include in the CSC2 definition the strategic level as well, e.g., over the program objective memorandum (POM) process in which CS plans need to be assessed, monitored, and controlled.

To date, operational planning has not sufficiently incorporated CSC2. It is challenging to do so for several reasons. CS planners usually do not have up-to-date and reliable CS resource information in a format that can be easily broken down for use by operators. Also, most logisticians are not trained in and do not participate in air campaign planning. Finally, operators are often unwilling to commit early on to plans (to the degree that they would put them in writing and pass them on to CS planners).

This work expands on the work of Leftwich et al.,[10] which presented initial concepts for guiding the development of a CSC2 operational architecture[11] for the AEF. When that work was started, the

An Operational Architecture for Combat Support Execution Planning and Control, Santa Monica, Calif.: RAND Corporation, MR-1536-AF, 2002.

[9] This report deals with the processes associated with CS execution planning and control. Often these processes have been referred to as the set of combat support command and control (CSC2) processes. We will use CSC2 to describe these processes in this report.

[10] Research at RAND defined an initial concept for a CS execution planning and control architecture. See Leftwich et al. (2002).

[11] An operational architecture, within the Department of Defense (DoD), is a description of tasks, operational elements, and information flows required to accomplish or support a DoD function or military operation. It describes the operational elements, assigned tasks and activities, and information flows required to support the warfighter. It defines the types of information exchanged, the frequency of exchange, which tasks and activities are supported by the information exchanges, and the nature of information exchanges in sufficient detail to ascertain specific interoperability requirements (Department of Defense, *C4ISR Framework Document Version 2.0,* December 18, 1997.) The Leftwich report and this report are not by themselves nor do they contain operational architectures per se. The results of these analyses are concepts that may guide the Air Force in developing and refining its CSC2 operational archi-

Air Force simply had no operational architecture for CSC2. Leftwich addressed the problem of CS not being integrated into operational planning, focusing mostly on the Commander of Air Force Forces (COMAFFOR) and Joint Forces Air Component Commander (JFACC) levels during strategic planning and contingency planning and execution. For example, during crisis action planning, Air Force operators had limited access to CS information to influence their decisions.[12] The Air Force began to implement Leftwich et al.'s recommendations but asked for further work. The current work introduces new concepts for Air Force involvement in the Planning, Programming, Budgeting, and Execution (PPBE) processes and provides further detail on CS contingency planning and execution processes associated with specific organizational nodes described in the earlier report. While Leftwich described some of the CSC2 information produced and passed between organizational nodes, this work goes further in depth and breadth, adding detail on those information flows and the decisions they drive.[13] We intend this study to guide the refinement of the Air Force's CSC2 operational architecture, specifically at the Headquarters Air Force Materiel Command (AFMC) and COMAFFOR A-Staff levels.[14]

We recognize that coalition support has been a key factor in recent U.S. military operations and that coordinating and integrating the CS of coalition partners has been a significant challenge. However, that issue was outside the scope of this report—which focuses on internal Air Force issues—and is not treated here.

---

tecture. For simplicity's sake, we will refer to our results as an operational architecture, although they are in fact concepts for such.

[12] During Operation Allied Force, the single CS person responsible for interaction in the operational planning group did not have a full depth of CS experience, information system links, or decision support tools to help facilitate interaction. For more detail on these and other issues, see Feinberg (2002).

[13] The CD-ROM enclosed with this document contains a library of dozens of proposed information products for several different organizational nodes.

[14] Rather than view the results of this study as a CSC2 operational architecture, which would promote the concept of a stovepiped, nonintegrated architecture, we address CS execution planning and control processes in the context of the larger Air Force C2 architecture.

Throughout this report, we refer to four different operational architectures: the AS-IS, TO-BE, the Evolving AS-IS, and the Expanded TO-BE. Leftwich et al. took as their starting point the existing operational architecture, calling it the AS-IS. The results of their research were assembled into what they called the future, or TO-BE. Because the actual operational architecture they observed has evolved since the original work (and continues to evolve)—due to Air Force–initiated changes and implementations of some of Leftwich et al.'s concepts—we refer to the current architecture that we took as the starting point for our analysis as the Evolving AS-IS. We analyzed the Evolving AS-IS architecture and built on some of Leftwich et al.'s more general architectural concepts. We refer to the assembly of our results as the Expanded TO-BE. Table 1.1 shows these four architectures, their place in this research, and a few distinguishing characteristics of each.

Our analysis leads us to believe that, to meet AEF demands, the Air Force needs to implement several changes to doctrine, policy, organizations, and information systems.

We present our expanded architectural concepts in three parts: *textual descriptions* of processes, organizations, and information products; detailed *process diagrams*; and graphical depictions of notional *information products*. This report contains complete textual descriptions in the main body, process maps in part in the body, and examples of information products in both the body and appendixes. The accompanying CD-ROM has electronic versions of the detailed process diagrams and the complete library of notional information products we have proposed.

In the next chapter, we discuss background to this research—the previous work on which this builds and a summary of the operational architecture as it has evolved over recent contingencies—and our analytic approach. In Chapter Three, we present the results of our analysis—the expanded CSC2 operational architecture. Following that, in Chapter Four, we describe shortcomings in the Evolving AS-IS operational architecture and propose changes to bridge those gaps. In Chapter Five, we summarize our work and offer several con-

Table 1.1
Current and Future Operational Architectures in Previous and Current Work

Architecture	Place in Research	Characteristics
AS-IS	Leftwich et al. starting point	Poor operations/CS integration Absence of feedback loops in planning and execution process Poor coordination of Air Force with joint community Absence of resource allocation mechanisms across competing theaters Inadequate understanding that CS refers to installations support
TO-BE	Leftwich et al. results	Enhanced Air Force CS doctrine and policy Standing CS organizations to conduct execution planning and control Operations and CS personnel trained on each other's C2 roles Appropriate information systems and decision support tools to translate CS resource levels and processes into operational capabilities or effects Specific focus on contingency planning at JFACC/major command (MAJCOM) levels
Evolving AS-IS	Current study starting point	Doctrine and policy reviewed, revisions to reflect the TO-BE CSC2 operational architecture Creation or adaptation of several organizations at global and COMAFFOR levels in line with TO-BE architecture
Expanded TO-BE	Current study results	Solutions to Evolving AS-IS shortfalls Adds detail on PPBE process to TO-BE Adds detail on Commodity Control Points (CCPs) to TO-BE

clusions. Appendix A contains more detailed descriptions of recent contingencies that are summarized in Chapter Two, and Appendix B contains two notional planning scenarios to illustrate features of the operational architecture. Two process diagrams have been placed in the enclosed CD-ROM to add further detail to descriptions of the operational architecture found in Chapter Three and Appendix B.

CHAPTER TWO

Background and Analytic Approach

We now discuss the objectives of CSC2 as derived from AEF goals, summarize the previous work on which ours expands, and trace the evolution of the Air Force's CSC2 operational architecture through several recent contingencies.

Objectives of CSC2

AEF operational needs drive CSC2 requirements, as shown in Table 2.1. Rapidly tailoring force packages requires that CS planners begin to generate support requirements based on desired operational effects alone. CS planners must coordinate closely with operators to estimate suitable force packages before such decisions are finalized. Early generation of CS requirements will contribute substantially to course of action (COA) assessment, focusing efforts on feasible COAs early in the planning process.

Rapid deployment requires that CS planners provide force beddown plans and assessments quickly. Generally, assessments begin before plans are finalized, and therefore the capabilities, capacities, and status of all potentially relevant airfields need to be available if quick turn assessments are to be made on the suitability of specific bases for receiving forces being contemplated. In addition, the status of in-theater resources should be continuously updated and effectively communicated to facilitate rapid Time-Phased Force and Deployment Data (TPFDD) development.

Quick employment and subsequent sustainment require that theater and global distribution, maintenance, and supply operations be rapidly configured and reconfigured to meet dynamic battlespace needs and that global prioritization and allocation of CS resources be rapidly shifted to areas of interest. Effectively allocating scarce resources requires that CS resource managers monitor resources in all theaters and prioritize and allocate resources in accordance with global operational needs. Finally, operational planners and resource managers should constantly monitor key performance parameters during execution and be able to adjust to changes in either CS performance or operational objectives.

Summary of Previous Work

CSC2 AS-IS Deficiencies

RAND's prior analysis of the Air Force's CSC2 process revealed important shortfalls in the AS-IS architecture, which can be grouped into the following five categories.

Table 2.1
CSC2 Functionality Required to Meet AEF Operational Goals

AEF Operational Need	CSC2 Requirements
Rapidly tailor force packages to achieve desired operational effects	Estimate CS requirements for suitable force package options; assess feasibility of alternative operational and support plans
	Identify and preplan potential operating locations
Deploy rapidly	Determine forward operating location (FOL) bed-down capabilities and capacities for force packages and facilitate rapid TPFDD development
Employ quickly	Configure distribution network rapidly to meet employment tasking and resupply needs
Shift to sustainment smoothly	Execute resupply plans and monitor performance
Allocate scarce resources to where they are most needed	Determine impacts of allocating scarce resources to various combatant commanders and prioritize allocation to users

Poor Integration of CS Input into Operational Planning. The conventional roles of the operations and CS communities often entail separate and relatively independent C2 activities. The traditional separation between the CS and operational planning communities hinders effective integration. At the same time, operators lack logistics or installation support training and hence tend not to consider the effect that support capabilities have on planned missions. An additional hindrance to incorporating CS input into operational planning is a lack of CS assessment capabilities and up-to-date and reliable CS resource information.

Absence of Feedback Loops and the Ability to Reconfigure the CS Infrastructure Dynamically. CS and operations activities must be continuously monitored for changes in performance and regulated to keep within planned objectives. Today, asset visibility is limited and in-transit visibility is poor. Thus, it is difficult to estimate current resource levels and future arrival times. CS feedback data—resource levels, rates of consumption, critical component removal rates, and critical process performance times such as repair times, munitions build-up times, in-transit times, infrastructure capacity, and site preparation times—may not be recorded routinely. Even when these data are available, they are typically the focus of planning and deployment rather than employment and sustainment. Because operations can change suddenly, these data must be continuously available throughout operations for operators and logisticians to make needed adjustments.

When monitoring reveals a mismatch between desired and actual resource or process performance levels, it may be difficult to find the source, particularly for activities supporting multiple theaters (such as depot repair), or multiple services (such as a Theater Distribution System [TDS] or construction priority). Discrepancies between desired and actual levels of support may arise from changes in CS performance or in operations. Assessment must be able to quickly address CS performance problems or changes and estimate CS requirements to meet changing operational objectives. With limited monitoring and performance assessment, it is hard to know when to intervene and adjust CS activities in real time.

Poor Coordination of CS Activities with the Joint Service Community. Ultimately, most CS (logistics and installations) activities entail coordination among the services and the joint service community. Nowhere is such coordination more important and troublesome than in transportation and distribution management. In principle, the distribution system can operate smoothly if those involved do their job and know their role; troubles can arise when the relative roles of the different contributors in an operation are not understood, expectations differ on anticipated performance, or priorities differ among the major players. Because the AEF relies on rapid distribution logistics and because CS depends on rapid and reliable transportation, rapid theater distribution systems should be developed that take full advantage of cooperation with the Army, Navy, joint service community, and allied or coalition forces (if applicable). If rapid resupply cannot be established, the Air Force may have to rethink lean policies and deploy with more resources to sustain operations, which would negatively affect deployment and employment timelines. Just as CS needs and capabilities must be communicated to operations planners, so too must they be communicated to, agreed on by, and resourced with other services, the joint service community, and allied or coalition organizations. Similarly, CS personnel should clearly define base capabilities to execute beddown plans and be prepared to provide those requirements to allied or coalition forces that may host Air Force units in a contingency.

Absence of Resource Allocation and Prioritization Mechanisms Across Competing Theaters. Resources planned for other regions must often be diverted to support a theater preparing for or engaged in a contingency. However, although the current process can allocate resources among units within a theater, it cannot formally allocate scarce resources across competing theater and joint task force demands or support analyses that should accompany requests for scarce resources. This type of assessment must be done before resources are reallocated so that high-level decisionmakers (up to and including the Joint Chiefs of Staff) can see the effect of their allocation decisions before the fact.

Inadequate Understanding That CS Refers Not Only to Logistics but Also to Installation Support. Attempts to incorporate CS inputs into operational planning faced not only the traditional separation between operations and logistics but also the separation between logistics and installation support. Logistics and their installation support counterparts grow from experience and training in two very different career paths. It is false to assume that in a contingency logisticians or installation support can rapidly become well versed in each other's diverse activities. Analysis of the CSC2 processes associated with the above three examples showed duplication of some activities when these CS functions acted independently but synergistic improvement when they teamed up. Thus, CS needs must be (1) managed by staff with adequate depth, experience, and rank and (2) integrated with CSC2 processes to focus the results.

CSC2 TO-BE Concepts and Operational Architecture for the Future

Leftwich et al. (2002) proposed a TO-BE CSC2 architecture that would enable the Air Force to meet its AEF operational goals. The architecture would enable the CS community to quickly estimate support requirements for force package options and assess the feasibility of operational and support plans. The architecture would permit quick determination of beddown needs and capabilities, facilitate rapid TPFDD development, and support development and configuration of a theater distribution network to meet Air Force employment timelines and resupply needs. The TO-BE architecture would facilitate development of resupply plans and monitor performance, determine impacts of allocating scarce resources to various combatant commanders, indicate when CS performance deviates from desired states, and facilitate the development and implementation of "get well" plans.

Shortcomings and Proposed Changes

Finally, the Leftwich et al. report offered the following recommendations to help the Air Force CS community move from the current architecture to the future concept described therein:

- *Summarizing and clarifying Air Force CS doctrine and policy.* The objectives and functions of execution planning and control must be recognized and codified in doctrine. The functions of concurrent development of plans among operators and CS personnel, assessment of plan feasibility, use of feedback loops to monitor CS performance against plans, and development of get-well planning need to be articulated and better understood.
- *Creating standing CS organizations to conduct execution planning and control.* The Air Force has supported one contingency after another for the last decade. Standing (permanent) organizations are needed to conduct CS functions and reduce turbulence and problems associated with the transition from supporting one contingency to reshaping support processes to meet the needs of another contingency.
- *Training operations and CS personnel on each other's C2 roles.* Understanding each other's responsibilities and methods can facilitate incorporation of both aspects into operational plans.
- *Fielding appropriate information systems and decision support tools to translate CS resource levels and processes into operational capabilities or effects.* This will improve understanding of CS constraints or value for an operational planning option.[1]

The Evolving Air Force CSC2 Operational Architecture

Here we visit the starting point for the current analysis—the Evolving AS-IS CSC2 operational architecture. The need for the level of CSC2 functionality described earlier in this chapter, as well as further insights into the needs of the CSC2 architecture, was revealed in Air Force operations in Operation Allied Force (OAF), Operation Enduring Freedom (OEF), and Operation Iraqi Freedom (OIF). The lessons from and shortcomings of the present architecture in these three conflicts provide useful insights for AEF CSC2 requirements.

[1] This section adapted from Leftwich et al. (2002).

The major observations (related to AEF operational needs) and corresponding CSC2 requirements for OAF, OEF, and OIF are summarized in Table 2.2. More detailed descriptions of each contingency are contained in Appendix A.

Analysis Approach

The analysis approach we used (illustrated in Figure 2.1) was similar to that of the previous work.

The first step was to use the previously defined[2] expected CSC2 functionality (shown in the outlined box). This was compiled from AEF operational needs, lessons from recent contingencies, discussions with subject-matter experts, and visits with many Air Force and joint organizations.[3]

We drew from the previously documented AS-IS architecture and continued to observe the Evolving AS-IS architecture, focusing on analyses of lessons learned from OEF and OIF, as well as recent Air Force initiatives—the Chief's Logistics Review, Spares Campaign, and Air Staff CSC2 implementation.

Using the desired functional characteristics and analysis of the AS-IS architecture, we expanded TO-BE concepts for the operational architecture using text and diagrams.[4] We also created a library of information products[5] produced by organizational nodes. The library includes information required to perform the tasks, the information sources, the producer of the products, and the recipients of the products.

---

[2] Leftwich et al. (2002).

[3] See Appendix A in Leftwich et al. (2002) for a list of interviewees.

[4] These descriptions and captions of diagrams are found in Chapter Four and Appendix B. The complete diagrams are found in the CD-ROM enclosed.

[5] Throughout this report, we use "product" and "information product" interchangeably to mean the output of an organizational activity in the architecture that in turn feeds a decision elsewhere in the system (e.g., plan feasibility analysis).

Table 2.2
CSC2 Requirements Revealed by Lessons from Recent Operations

Conflict	Observations	CSC2 Requirements
Operation Allied Force	Slow and difficult transition from peacetime to wartime operations	Identify permanent organizations that will perform critical CS tasks continuously during peace and war Expand Air Force involvement in theater distribution system planning and execution
	Poor interface between operations and CSC2	Include CS input in initial planning Translate CS information into operational capability
	Inability to react quickly to changes in the operational plans	Provide real-time visibility of theater and global resources Rapidly reconfigure CS infrastructure
	Insufficient and inadequate CSC2 policy and procedures; information systems; training; and education	Develop and formalize doctrine and policy; information systems; and training programs
Operation Enduring Freedom	Slow and difficult transition from peacetime to wartime operations	Identify permanent organizations that will perform critical CS tasks continuously during peace and war
	Confusion of responsibilities; duplication of effort	
	CS resources stretched thin	Include CS input in initial planning
	Incomplete information feedback loops	Complete CS and operations feedback loops
	CSC2 organizations exhibited flexibility and responsiveness in Centralized Intermediate Repair Facility (CIRF) operations	Continue using and standardizing CIRF operations Improve CONUS support location (CSL) linkages
Operation Iraqi Freedom	Eased transition to war	Initially adopt CSC2 architecture Better understand organizational responsibilities
	CS input in initial planning	Allow adequate time before combat operations Employ good logistics supportability analysis

Table 2.2—Continued

Conflict	Observations	CSC2 Requirement
Operation Iraqi Freedom (cont.)	CSC2 organizations exhibited flexibility and responsiveness in CIRF operations	Continue using and standardizing CIRF operations Improve CSL linkages
	Deployment planning process and information system slow to react to changes in operational plans	Improve deployment planning process and information systems

Figure 2.1
Analysis Approach

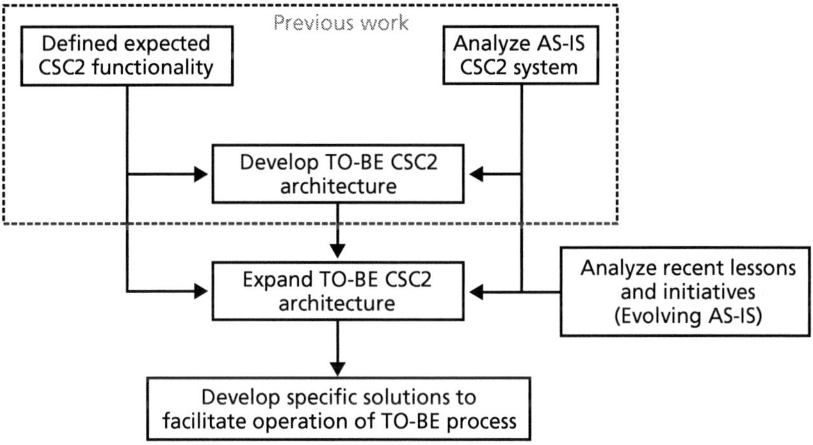

RAND *MG316-2.1*

The updated AS-IS architecture analysis was then compared with the AEF CSC2 needs to produce our main results: AS-IS shortcomings and changes necessary for the Expanded TO-BE architecture. These shortcomings were broadly grouped according to the type of modification ("solution") that would address them. For each category, we summarize the recommendations of the previous analysis, evaluate recent Air Force progress in addressing them, and discuss remaining shortcomings and how each hinders efficient CSC2. We then propose solutions aimed at resolving the shortcomings. The solutions are designed to facilitate, enhance, and refocus the operation of the CSC2 architecture to be in line with the desired functionality and Expanded TO-BE concepts.

CHAPTER THREE

Expanded Combat Support Execution Planning and Control Architecture for the Future

This chapter comprises two major elements: key aspects of the TO-BE operational architecture from Leftwich et al.'s work that laid the groundwork for this analysis, and the Expanded TO-BE concepts that this work contributes.

Leftwich et al. described the TO-BE processes at three levels: high, medium, and detailed. The high level was a generic closed-loop planning and execution process. The medium level applied this generic process to each phase of operations and used it to highlight shortfalls in the AS-IS architecture. The detailed level was communicated in elaborate process maps. Also, Leftwich et al. recommended the formation and formalization of three kinds of standing CS organizations.

In this chapter's first section, we present condensed descriptions of the high-level (generic) planning and execution process and of the three standing CS organizations. With these we describe the foundation of Leftwich et al.'s work on which this work expands. In the second section, we present the heart of this analysis: textual descriptions and process diagrams of Expanded TO-BE architectural concepts. We introduce new concepts for organizational functions and information in the PPBE system, and we add detail to Leftwich et al.'s work on contingency execution planning and control.[1] We do

[1] Not included in this chapter are notional examples of the expanded architecture in deliberate planning (not included in Leftwich et al., 2002) and contingency execution planning and control (expanded from Leftwich et al.). The information product library on the enclosed CD-ROM is also a new addition since the Leftwich et al. work.

this in four subsections: planning, programming, budgeting, and execution; crisis action planning; deployment; and employment and sustainment.

Process and Organizational Concepts from Previous Work

High-Level TO-BE Process

The TO-BE concepts described in this and previous work integrate operational and CS planning in a closed-loop environment, providing feedback on performance and resources. Figure 3.1 illustrates these concepts in a process template that can be applied through all phases of an operation from readiness, planning, deployment, employment, and sustainment to redeployment and reconstitution. The figure centers on integrated operations and CS planning and incorporates activities for continually monitoring and adjusting performance.

Some elements of the process, in the shaded gray area of Figure 3.1, take place in planning for operations and should be accomplished as concurrently as possible. A key element of planning and execution in the process template is the feedback loop that determines how well processes are expected to perform (during planning) or are performing (during execution) and warns of potential failure. It is this feedback loop that tells the logistics and installations support planners to act when the CS plan and infrastructure should be reconfigured to meet dynamic operational requirements, during both planning and execution. The CS organizations will need to be flexible and adaptive to make changes in execution in a timely manner.

The feedback loop not only drives changes in the CS plan but might call for a shift in the operational plan. For CS processes to provide timely feedback to the operators, they must be tightly coupled with their planning and execution processes and information systems. Feedback might include notification of missions that cannot be performed because of CS limitations.

Figure 3.1
High-Level TO-BE Process

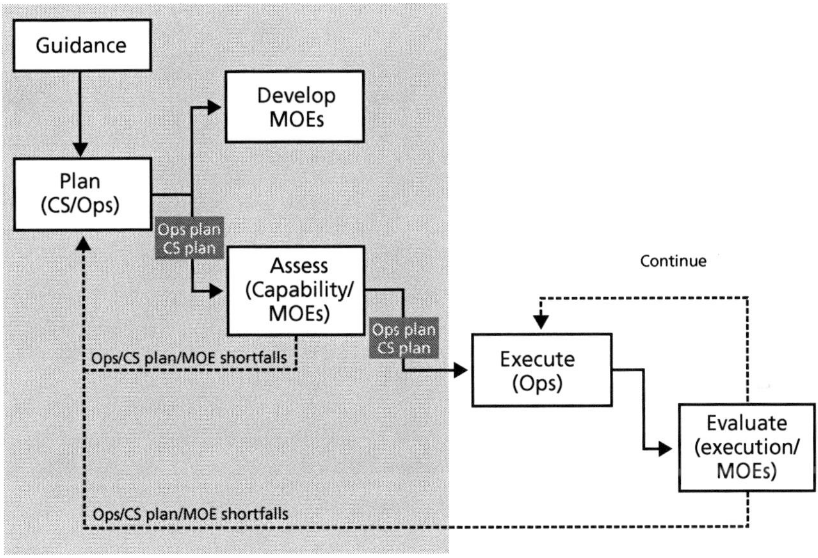

RAND *MG316-3.1*

Standing CS Organizations

One of the key nodes in the TO-BE CSC2 operational architecture is the subset of the COMAFFOR A-Staff performing CS and logistics functions.[2] (In Figure 3.2, this organizational node is portrayed by activities in the third and fourth horizontal bands from the top.) The COMAFFOR A-Staff can have a component that may go forward to the area of responsibility (AOR) with the COMAFFOR. It can also have a portion of the A-Staff that remains in the rear and serves A-Staff functions in a reachback fashion. Leftwich et al. (2002) referred to the CS and logistics subset of the COMAFFOR A-Staff

---

[2] We consider CS and logistics-related functions to be A-4, A-6, and A-7 staffs. This report speaks to their functions, not operations-related functions such as A-3 or A-5.

Figure 3.2
Planning, Programming, Budgeting, and Execution Process

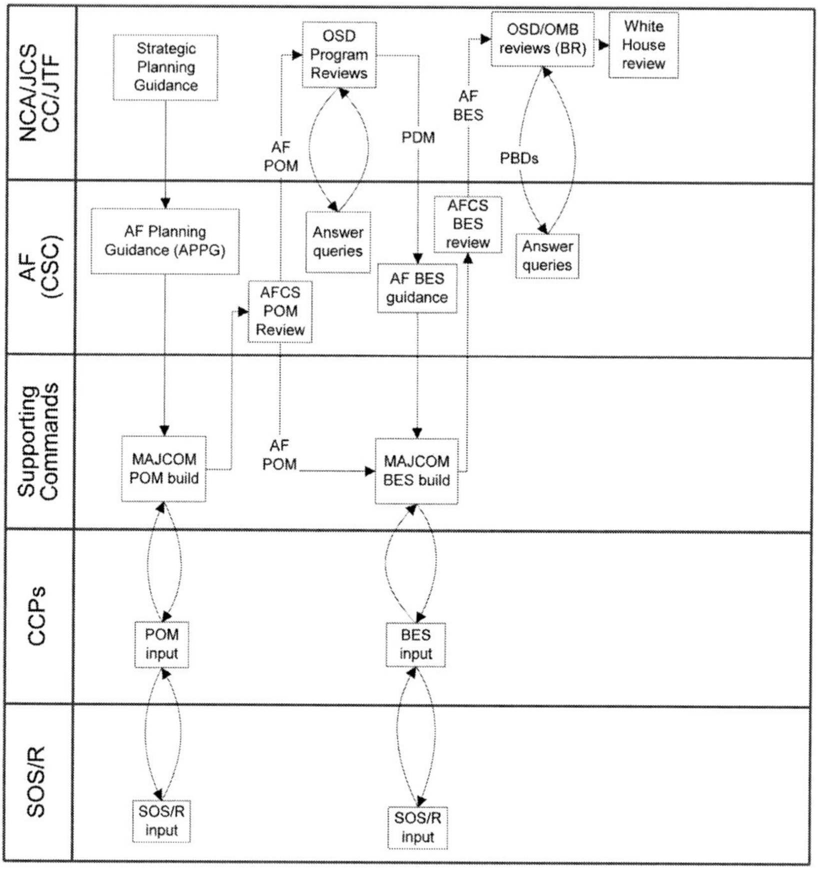

RAND *MG316-3.2*

Rear as the Operations Support Center (OSC).[3] The OSC will act as regional hub for monitoring, prioritizing, and allocating theater-level CS resources and be responsible for mission support, base infrastructure support, and establishing movement requirements within

[3] The distribution of A-Staff functions between forward and rear nodes may vary between COMAFFORs. Improved communications capability will enable increasingly more functions to be performed in the rear.

the theater. In general, the OSC will be the theater integrator for commodities managed by commodity control points. The OSC will have complete visibility of theater resources and the authority to reconfigure them. It will receive commodity-specific information from CCPs and make integrated capability assessments (both sortie production and base) and report those assessments to the A-4 Forward who will lead the CS element in the air operations center (AOC). In this role, the OSC will make allocation decisions in the face of competing demands for resources. Finally, it must work closely with the joint service forces community to ensure that resources are allocated in accordance with global priorities.

CCPs (depicted in Figure 3.2 as the second horizontal band from the bottom) will manage the supply of resources to the major commands (MAJCOMs)—essential for the distribution of such critical resources as munitions and spares. The CCP will monitor resource inventory levels, locations, and movement information and use these data to assess contractor and organic capabilities to meet throughput requirements. As an integrator, the CCP will bring together information from across traditional stovepipes to develop and improve feasibility and executability of plans, cost estimates, and budgets as well as to centralize buy and repair authority. It will also have the ability to allocate and reallocate cost authority during execution when priorities or conditions change.

In addition to the CCPs and the COMAFFOR A-Staff, a global Combat Support Center (CSC) is a critical node. This node, depicted in Figure 3.2 as the second horizontal band from the top, should assess weapon system capabilities and have responsibility for providing integrated weapon system assessments across commodities in both peacetime and wartime. With a global view, this Air Force–level node could integrate assessments that support allocation decisions when multiple theaters are competing for the same resources and could serve as the Air Force voice to the Joint Staff in any arbitration across services.

The CSC could be a self-contained organization on the Air Staff with analytic capabilities or a virtual organization with analysis cells collocated with the logistics support centers (LSCs).[4] A virtual organization would probably have a reduced manpower burden, since analysis cells could be carved out of existing LSCs, but it might add an information system requirement to aid communication with the Air Staff. A stand-alone Air Staff organization might be more responsive to immediate analytic needs but would require adding new manpower slots.

Expanded TO-BE Architectural Concepts

We will now discuss the Expanded TO-BE architecture and its application in support of planning and execution. We start with a general description of the PPBE system and then treat each component separately. Within these descriptions, we will differentiate between current architectural concepts and expanded concepts.

General Description of the Planning, Programming, Budgeting, and Execution System

The Department of Defense (DoD) PPBE system is a comprehensive process that takes four years to complete just one cycle. (This process is depicted graphically in Figure 3.2.) Although it is advertised as a biennial process (with only an update of the budget in the second year), in practice, the dynamics of DoD require a compilation every year. This means that in any one year, the process has four overlapping budget positions in various stages of development. The PPBE system transforms military plans into programs, creating the Future Years Defense Program (FYDP) for two budget years and five outyears and then providing the cost of those programs to Congress for its approval and funding. The FYDP is a database tool that keeps DoD management informed of what has been accomplished in the recent past and what is slated for funding and accomplishment in the short to medium term. The FYDP represents how DoD money will

---

[4] At the time this analysis was performed, these were called regional supply squadrons (RSSs). Each MAJCOM had its own RSS. Now, there are two LSCs—one for combat forces and one for mobility forces. Elsewhere in the document, we use the term LSC.

be obligated in future years based on plans and obligations approved by the Office of the Secretary of Defense (OSD) during previous PPBE cycles.

In the fall of the first year of the PPBE cycle, Headquarters Air Force (HQ-AF) uses OSD's Strategic Planning Guidance to create the Air Force Planning and Programming Guidance (APPG), which guides the MAJCOMs in developing their inputs for the POM. In the second year, OSD provides the Joint Programming Guidance, which defines the fiscal constraints for the POM. When the MAJCOMs submit their POM inputs, they are reviewed by a series of mission panels that look carefully at the inputs and resolve difficult issues at increasingly higher levels largely from the warfighters' point of view. For example, the Agile Combat Support Panel, chaired by the Air Combat Command (ACC), reviews and adjudicates issues for the spares budgets.

While this is happening, the Air Force turns its POM into a Budget Estimate Submission (BES). This is when pricing factors are updated, earlier decisions are examined in light of recent changes, and changes are negotiated with OSD. Panels at HQ-AF review these inputs again, resolving problems by moving up the chain of command as problems get tougher and eventually producing a fully coordinated budget. After review by OSD and Office of Management and Budget, the DoD budget is sent to the President for approval and adjustment. In January or February of the third year, the President sends it to Congress. The following year, Congress revises, approves, and appropriates a budget that allows DoD to begin executing the budget that resulted from this one four-year budget cycle.[5]

Planning

Recalling Figure 3.1, the activities portrayed there can be applied to deliberate planning, as depicted in Figure 3.3. Here, the vertical hier-

---

[5] Summary borrowed and revised from Frank Camm and Leslie Lewis, *Effective Treatment of Logistics Resource Issues in the Air Force Planning, Programming, and Budgeting System (PPBS) Process*, Santa Monica, Calif.: RAND Corporation, MR-1611-AF, 2003.

Figure 3.3
Planning Process

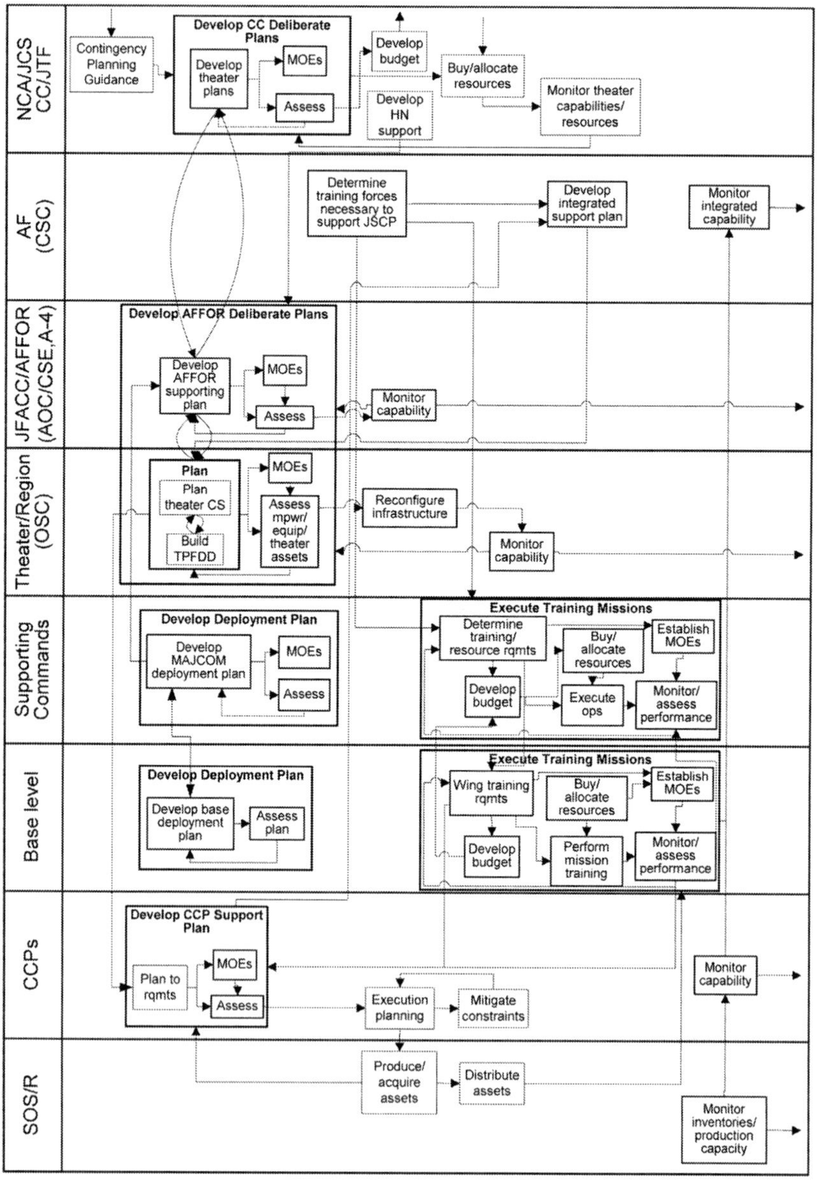

RAND *MG316-3.3*

archy corresponds more to information flow than to authority. Below, we focus on the CS aspects of each process and identify the interactions with operators. In Leftwich et al. (2002), the description of planning included estimating CS resource needs, operational planning products, infrastructure configuration, capability assessment, and plan feasibility assessment. We add no new concepts to that description, but we do expand the descriptions of feasibility assessment, infrastructure configuration, and capability assessment. The descriptions below do not match current processes. They are suggestions for how the future system ought to function.

During deliberate planning, Air Force planners will go through an iterative planning process similar to that portrayed in Figure 3.3 (the following description corresponds to the third and fourth bands of the figure). Air Force planners will produce a portfolio of operational plans (could be dozens or hundreds) with associated deployment requirements (e.g., type and number of aircraft, sortie rates, beddown locations). From this, COMAFFOR CS planners (in the OSC) will produce corresponding requirements TPFDDs (what manpower and equipment, expressed as unit type codes [UTCs], will be required at each base), which will then be sourced and time phased. The CS portions of these plans will then be assessed for feasibility across each kind of requirement—manpower, equipment,[6] sustainment, and transportation—by its corresponding organizational node. To check for manpower and equipment feasibility, CS planners in the OSC will draw from the requirements TPFDD. They will use their visibility of AEF capabilities to determine which portion of Air Force forces each plan requires, whether the current AEF construct could meet the requirements for a sustained period or whether AEF "rules" would have to be broken (or whether the Guard or Reserves must be called up), and what capabilities might be left over if such an operation were conducted as planned.

Another layer of feasibility that must be tested is that of the regional or theater resources. OSCs, serving as COMAFFOR A-4

[6] By equipment we mean non-consumables, both unit equipment and WRM, e.g., aerospace ground equipment, FMSE, vehicles.

Rear and Forward nodes, maintain visibility over centralized intermediate repair facilities and beddown sites (forward operating locations/ forward support locations [FSLs]) in their theater. During planning, the rear node will determine Centralized Intermediate Repair Facility (CIRF) activity levels necessary to support each plan, including manpower and equipment needed to augment peacetime levels. Using its visibility of potential beddown sites, the rear node will assess the ability of each location in the plan to support the specified forces and what additional equipment and manpower is needed at each. It might also propose alternative locations that could better support the specified forces.

When these planning options are being developed and updated, the CSC will pass on draft options to the CCPs to get updated materiel data. The CCPs will then mobilize their information to determine feasibility (and resource trade-offs if plans are not feasible) and to identify any constraints. (For example: Will current spares production support flying programs and readiness for the contingency? Will munitions stores last through the expected duration of the contingency?) The CSC will compile data from all the CCPs to get a complete picture for each aircraft type. The CSC also analyzes the impact of these resource requirements on other military requirements. (For example: Will an aggressive plan in Central Air Forces [CENTAF] using a large number of Joint Direct Attack Munitions [JDAMs] require swinging stocks from PACAF [Pacific Air Forces] stores? What impact does this have if military action is required in Korea?) Key commodities to be analyzed are spare parts, munitions, basic expeditionary airfield resources (BEAR), fuels mobility support equipment (FMSE) and fuel, and bulk commodities (food, water, construction materials).

Finally, transportation feasibility must be assessed to see whether, given allocated lift resources and constraints, combatant commander timelines can be met. Here, planners must have a proposed execution TPFDD with time phasing and sourcing. While the United States Transportation Command (USTRANSCOM) would accomplish this task, an OSC would estimate Air Force deployment and sustainment movement requirements.

This whole process is iterated with modifications made to operations or CS plans, until a feasible plan is found.[7]

The early inclusion of OSCs and CCPs builds concurrency into the planning and identifies serious limitations in time to adjust plans when necessary, which is critical when many plans must be created. Involving these nodes early will produce more realistic options for the warfighter because serious showstoppers will be identified, alternatives suggested, and corrective action taken before execution begins. Support plans have always been part of military planning, but their feasibility is verified only at an aggregate level, and many broad assumptions are made. During execution, when weapon systems have not been available to perform their planned missions, ad hoc solutions have had to be implemented. In the expanded architecture, support plans are analyzed early to determine whether they are executable, using fewer assumptions, more facts. This early involvement may also reveal to planners some issues or problems of resource availability that were not previously known but that now may be managed proactively.

Infrastructure Configuration. Comprehensive capabilities-based portfolio planning generates global requirements that will necessitate changes in the CS network to best prepare for a range of scenarios. CS configuration must be done at a global, strategic level to prepare the Air Force to meet future demands, including infrastructure, materiel, and personnel.

Analysis at the Air Staff level will be done to assess global requirements for all resources and, for some, how they will be postured among FSLs and FOLs.[8] The actions that follow this analysis will best prepare the Air Force for an uncertain future. Each CCP will have input into the process, but the analysis will be done in an inte-

---

[7] Paul K. Davis, *Analytic Architecture for Capabilities-Based Planning, Mission-System Analysis, and Transformation*, Santa Monica, Calif.: RAND Corporation, MR-1513-OSD, 2002, describes in more detail a proposed methodology for capabilities-based planning.

[8] This analysis ought to be done often enough to keep pace with changing Secretary of Defense priorities (e.g., Strategic Planning Guidance) and plans, but infrequently enough to make it financially feasible.

grated fashion so that resources are balanced (i.e., different resources are not postured optimally for different kinds of scenarios). For example, an output of this analysis will be how many total BEAR assets are needed and in which FSLs they should be stored. This will take into account the range of scenarios, FSL and FOL maximum on ground and throughput, travel times, and transportation limitations (e.g., ports in third world countries that will accept only high draft water vessels). Another very different example is that of the spares CCP. This node will take the multiple contingency requirements or readiness spares packages and spares, combined with peacetime requirements to support training, and determine the required production levels and cost and performance trade-offs. (More on CCP interaction with the budgeting process will follow.)

CCPs will manage the actual purchase or repair of assets to support this global asset posturing. New contracts will be let if necessary. OSCs will be the data repository of FOL/FSL capabilities and limitations and will monitor the execution of any infrastructure additions in their AOR in support of the global posturing.

For each contingency plan and for steady-state requirements, planners will also reevaluate planning factors, establishing new targets for CS measures of effectiveness to meet the contingency plans. These factors include parking capacity of aircraft ramps, potential fuel consumption versus available fuel storage and distribution capacities, critical water and power capacities, expected removal rates for reparables, expected repair times for commodities through the various repair facilities, expected response times at various points within the distribution network, expected munitions expenditure rates, and attrition rates due to enemy action. These planning factors become critical inputs to the decision support tools that provide the "look-ahead" capability that enables CS to be proactive.

Capability Assessment and Reporting. An important feature of the Expanded TO-BE architecture is a comprehensive capability assessment and reporting system. Planning for contingencies and configuring the CS infrastructure and resources will be done, but to complete the feedback loop (as depicted in Figure 3.3), CS capabilities must be measured and compared with planned levels. Without accurate knowledge of its resource capabilities, the Air Force will be unable to mitigate shortfalls and may be caught unawares when a contingency arises.

Periodically, organizational nodes will assess their performance in supporting operations and project their ability to meet Secretary of Defense requirements.[9] These projections include bottlenecks in achieving goals, time to correct deficiencies, and new or reallocated monies that could alleviate the problems. The CSC receives all CCP assessments and projections and presents periodic updates to the Air Staff, along with problems and proposed get-well plans. For instance, the spares CCP will periodically determine when projected operational performance deviates from plans, determine the cause(s) of deviation, and test alternative get-well approaches, including reallocating money. The munitions CCP will check levels of its munitions (especially important if current contingency operations are depleting stores of critical munitions) and their capabilities, and the throughput of its munitions dumps, especially if any infrastructure changes have been made since the plans were formed. CCPs will centrally manage other war reserve materiel (WRM) (e.g., BEAR, vehicles) and will provide global visibility over each resource. These capabilities will be available to a wide audience of Air Force leaders. Detailed information will be entered by CCP cells into a common electronic workspace (akin to the Air Force Portal), and much of the aggregation and capability assessment and reporting will be done by automated programs.[10]

Each OSC will periodically assess the capability of its resources, and each will measure the peacetime and wartime projected capability to support operations for CIRFs and base infrastructure. For instance, an engine CIRF will measure and report its ability to support sorties at levels maintainable during peace (i.e., for training only) and with

---

[9] Capability assessment cycles should be created and tailored for each capability—often enough to keep up with the pace of changes, but not so frequently that they become time or cost prohibitive.

[10] We refer to this portal or system as the "common electronic workspace" or simply the "electronic workspace" for the remainder of the report.

plus-ups for war (including the manpower and equipment augmentation necessary to achieve certain capability levels). Each OSC will report on the state of the infrastructure of each base under its watch (e.g., aircraft supportable, maximum on ground). The OSC will report which capabilities are operational and which are required to ramp up to "full" operations.

This more precise capability assessment and reporting system presents new security risks. Once the Air Force analyzes and expresses its capabilities with this level of fidelity (e.g., quantifying how many sorties will be lost at FOLs if a backlog occurs at a particular transportation hub), it then possesses information valuable to its own planners and to its enemies. This information is thus a two-edged sword—added capability comes with added risk. The system for capabilities analysis and reporting must have security adequate to protect this information.

A notional example of deliberate planning can be found in Appendix B. We now discuss programming and budgeting, next in the PPBE process. Programming, budgeting, and execution are entirely new additions to the operational architecture; none was covered in Leftwich et al.'s work (2002).

Programming and Budgeting

Because the CCPs are key resource management nodes, they will play a major role in Air Force cost estimation. The deliberate planning described above will establish requirements for many commodities directly, and each CCP will perform cost estimates of global requirements for its commodity to feed the Air Force POM process.

Spare parts, however, have a unique process. As part of a logistics planning activity that operates in the background, behind the PPBE planning process, all MAJCOMs participate in the Air Force Cost Analysis Improvement Group process. This process seeks to define the Air Force–wide "fully funded requirement" for depot-level reparables (DLRs) (and some other important expenditures) during the year of execution. This process is entirely advisory and can influence actual funding proposals for DLRs, made in the operational commands, only by providing informational input.[11] Here, the spares CCP will produce analysis to inform the Spares Requirements Review Board (SRRB) process. It will take Air Force–stated weapon system availability (WSA) targets set for contingency and training requirements and combine them with logistical planning factors. It will communicate trade-offs, such as WSA achievable, in terms of dollars spent on spares for each mission design series (MDS). Actual usage data combined with prices charged by suppliers and sources of repair, as well as trend data showing historical price behavior, will be contrasted with the model outputs.[12]

Execution

When it comes to day-to-day budget execution, the CCP is the focal point, supporting units and maintaining a healthy supply chain (portrayed in Figure 3.4). The CCP will direct the flow of information, products, services, and financial management for its commodity. It will track the constraints in capacity, expertise, funding, and human resources routinely providing data and analysis to the common electronic workspace. In addition, the CCP will operate within a network of collaborative relationships—some formal, others informal. For spares, for instance, sister organizations at AFMC, such as the Warfighter Sustainment Division, will provide resident maintenance, supply, munitions, transportation, and logistics plans expertise when needed. The CSC at the Air Staff will provide liaison to the joint community and be the conduit for Air Force plans, policies, and priorities. The logistics support communities at the MAJCOMs will serve as the CCPs' direct link to the warfighters.

---

[11] Summary borrowed and revised from Camm and Lewis (2003). Some changes have recently been made at the DoD level in this process. These change some of the high-level planning documents given as instruction to the services, as well as the timing of some of the PPBE process, but do not change how internal Air Force processes work.

[12] This concept of the CCPs informing the SRRB process is a major feature of the expanded architecture.

Figure 3.4
Executing the Program

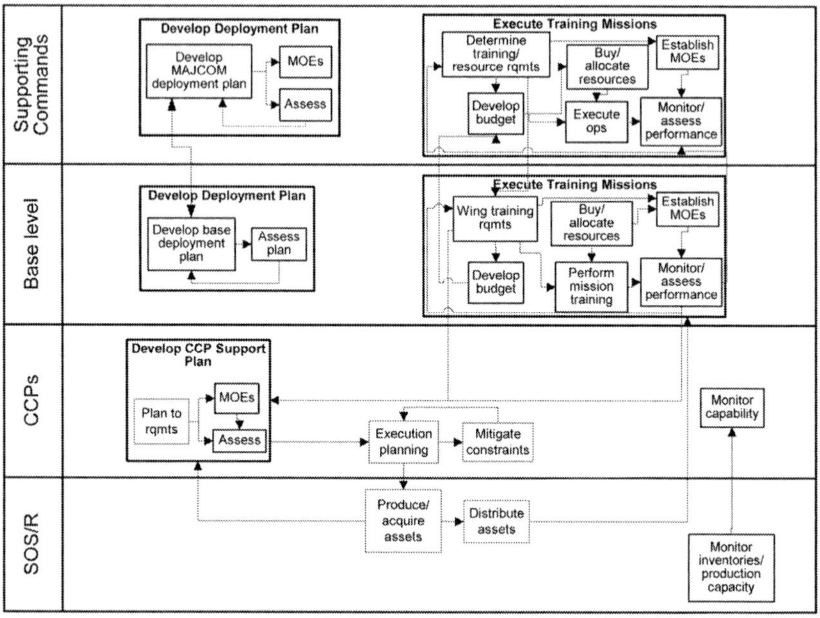

RAND *MG316-3.4*

Supply chain managers,[13] who are key players in the CCP team, will routinely provide information to the electronic workspace, including tracking of suppliers' financial performance and production capacities,[14] forecasting demand, determining lead times, determining transportation and delivery routes, and developing action plans when problems arise. They will also develop contract terms and conditions for sources of supply and/or repair that will enable the surge actions and increased data requirements needed during wartime. They will

[13] The spares CCP, in addition to commodity supply chain managers, will have weapon system supply chain managers (WS-SCMs) who focus on integrating information for a particular MDS.

[14] Recent contingencies highlighted the importance of having visibility into the production capacities of suppliers of critical items (e.g., precision munitions), especially in the face of dynamic requirements.

adopt a supplier's point of view, forecasting and managing downstream problems. Similar to their commercial counterparts, they will perform analyses and choose suppliers based on attractiveness: What is the supplier's performance history (quality, delivery record, price)? How well does the supplier manage technology changes? Is the supplier financially stable? The Air Force needs high-performing suppliers, and the CCP team will perform rigorous evaluation and selection and then develop long-term collaborative relationships with them. With guidance from HQ-AF, the CCP team will develop enterprise-wide strategies for those goods aligned with strategic goals, then implement the plans and develop instruments to measure outcomes.

The CCP will also manage the allocation of cost authority across weapon systems to the supply chain manager for both procurement and maintenance. The basis of the weapon system buy lists and repair actions will initially be the full requirement, but when cost authority is released and turns out to be less than budgeted,[15] the CCP will reallocate it according to MAJCOM/HQ-AF priorities.[16] Revised buy lists and repair actions will then be generated and executed. Whether or not full cost authority is received, closed loop feedback mechanisms will be developed to track the execution of the cost authority to determine whether operational goals are being met. In the event that the feedback results show that execution is not working well and analysis determines that another allocation could better support availability goals, the CCP will have the ability to reallocate cost authority.

In summary, deliberate planning in the expanded architecture includes major roles for OSCs: testing portfolios of plans for feasibility across multiple functions, measuring and communicating CS resource capabilities, and maintaining information security. New concepts for programming, budgeting, and execution primarily involve the CCPs: estimating budget requirements, planning pro-

[15] Many Air Force weapon systems are chronically underfunded and cannot meet current operational plans. This problem requires very effective CS execution planning and control to enable systems to be maintained with whatever resources are available.

[16] This concept of the CCP reallocating less than full funding is another new concept in the expanded architecture.

curement and repair (on a months-to-years time frame), and adjusting resource allocation to meet changing requirements or to prepare for uncertainties.

Crisis Action Planning

The previous section described planning, programming, budgeting, and execution. These processes occur constantly as part of the DoD bureaucratic system. The next three phases we discuss—crisis action planning, deployment, and employment and sustainment—occur in response to contingencies. We first discuss crisis action planning.

The crisis action planning process will essentially mirror deliberate planning, with a few exceptions. If and when a contingency requiring military force arises, all the strategic or long-term planning and execution activities continue in the background. Deliberate plans are made, budgets are created and executed, resources are acquired and allocated, units train, and various organizations monitor Air Force performance and capabilities.

At the outset of the crisis action planning process, as the combatant commander and the service components explore COA options, COMAFFOR planners access the capability assessments OSCs and CCPs have been periodically producing, reporting, and archiving in the electronic workspace. These assessments (e.g., beddown locations, WRM stocks, spares surge capability) form the "resource landscape" of the Air Force and inform the COA selection process.

Once the combatant commander selects a COA, operators create an Operational Parameters Template (OPT) to jump-start CS planning. An OPT (depicted in Figure 3.5) is a single document, containing estimated operational requirements (MDS, sortie, beddown), from which CS planners will begin their CS planning.[17] It contains much of the same information as a wartime aircraft activity report, except it is tailored to the current situation. Without this sin-

[17] The production and use of an operational parameters template are key features of the expanded architecture; neither is currently done.

Figure 3.5
Operational Planning Template

Operational Parameters Template					
Base	**Unit**	**Sorties/Day**	**ASD**	**Start**	**End**
Base A	123 TFS	1.1	2.8	0001	0010
		0.8	3.2	0011	0030
		0.6	3.2	0031	0060
	456 TFS	1.4	2.0	0001	0006
		1.2	3.5	0007	0010
		1.0	3.0	0010	0020
	...	...	...	...	...
	...	...	...	...	...

RAND *MG316-3.5*

gle planning document, or "sheet of music," CS planners will have to make their own guesses and assumptions about operational characteristics. The CS planning process would immediately be off to an uncoordinated, haphazard start.

This template is passed to COMAFFOR Forward and Rear nodes. Forward planners do more situation-specific beddown planning to feed air campaign planning. Functional experts in the COMAFFOR Rear works from this information to create functional slices of the TPFDD. After the pieces are assembled, OSC planners source the forces.

As in deliberate planning, CCPs and OSCs perform feasibility assessments (e.g., manpower, equipment, sustainment, transportation) on plans, albeit more quickly. If there are obstacles or bottlenecks to achieving combatant commander goals, then these nodes perform get-well planning, which may entail shifting resources from other weapon systems, COMAFFORs, or theaters. Whatever the case, the AOC and key CSC2 nodes coordinate to meet the requirements.

Also similar to deliberate planning, Air Force global CS infrastructure must be tailored and reconfigured to meet contingency requirements. This includes opening FOLs, augmenting FSLs, and deploying personnel and equipment to operating locations. The OSC Rear combines CCP assessments to feed air campaign planning, and the spares CCP performs a spares feasibility assessment. If an air logistics center (ALC) surge is needed, the ALC may have to sacrifice training or readiness of other weapon systems to make readiness spares packages and meet combatant commander goals.

Deployment

The COMAFFOR is a major focal point of information flow and decision for the deployment process. The COMAFFOR Forward, with specific knowledge of its AOR, performs beddown planning. The rear element handles TPFDD development, and the forward element then monitors force reception. Deploying units have visibility of the TPFDD and when and what they must deploy. While USTRANSCOM plans the lift, the COMAFFOR Rear manages the connection between USTRANSCOM and deploying units, ensuring communication and solving problems when they exist (e.g., unit unable to deploy on time). The COMAFFOR Forward communicates initial force reception information on each FOL to the AOC. As deployment continues, even after operations have started, the COMAFFOR Forward continues to monitor and report force reception and force reception information to the AOC combat support element. The COMAFFOR Forward liaisons with the COMAFFOR Rear to help solve beddown problems as they occur. This process is depicted in Figure 3.6 (middle left).

Employment/Sustainment

The major CS activity during employment and sustainment (depicted in Figure 3.6, right side) is monitoring performance. High-level met-

Figure 3.6
Deployment and Employment/Sustainment Processes

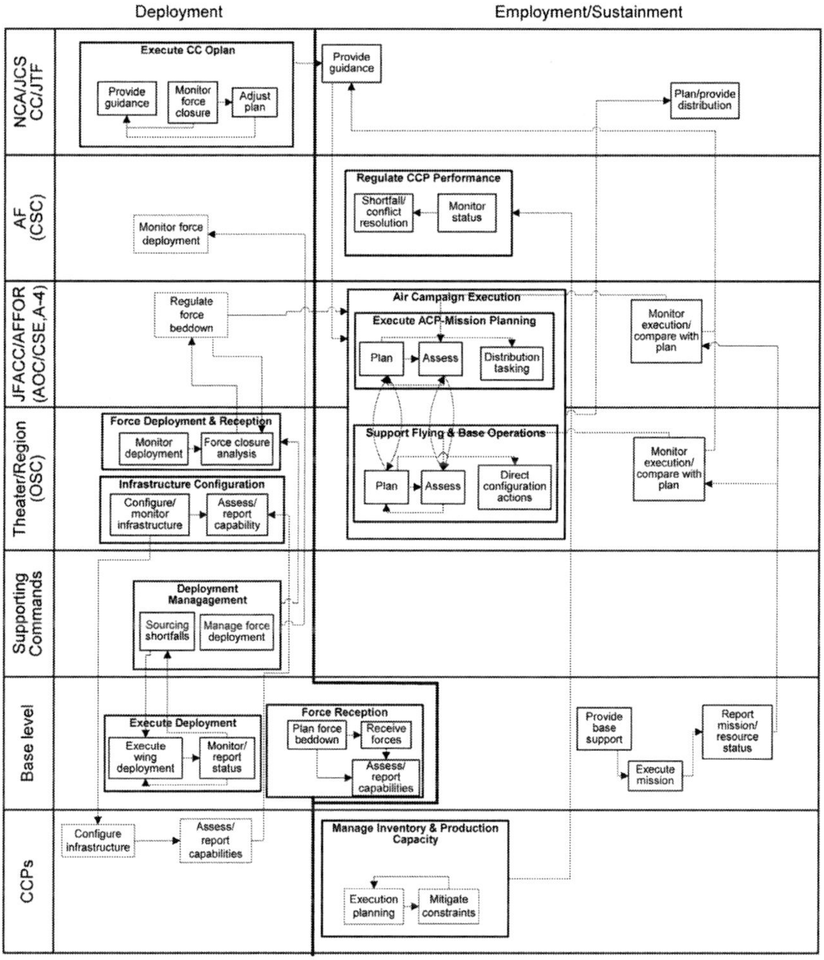

RAND *MG316-3.6*

rics are monitored, which drives allocation and planning decisions. Low-level metrics are monitored and feed into the reporting system. FOLs monitor such metrics as spares levels, repair times, munitions levels, and infrastructure condition and capabilities. The OSC monitors and integrates these metrics and plans get-well actions when actual performance deviates from planned levels. The COMAFFOR Rear monitors the TDS and plans adjustments as necessary.

The CSC monitors force-level metrics and arbitrates when resource decisions cannot be made within a theater. CCPs supply commodity resources, monitor processes, and report performance. For munitions, for example, FOLs monitor stocks and report their status. The OSC observes FOL and WRM stock levels and CCP information, and it reports status and problems to the AOC for planning. Similar integration is done for spares. FOLs and CIRFs monitor stock levels and repair times. The OSC integrates these with delivery times and works with the LSCs (formerly RSSs) and CCPs to ensure adequate spares support.

A notional example of contingency planning and execution can be found in Appendix B.

CHAPTER FOUR

Current Progress Toward Implementing the TO-BE Architecture and Recommendations for Meeting the Expanded Architecture

The Expanded TO-BE concepts discussed in Chapter Three have an execution planning and control process designed around the needs of the AEF—that is, it is operationally relevant, rapid, and responsive. The Air Force has been implementing many changes recently that correspond to the TO-BE architecture. Process improvements are core to the evolution of an execution planning and control architecture capable of ensuring support to the AEF. Several "enabling mechanisms," including doctrine and policy, organizational responsibilities, training and education, and information systems, are in the process of being modified to meet the intent of the TO-BE architecture.

In this chapter, we identify the recommendations from the TO-BE architecture, discuss Air Force progression toward realizing these concepts, and identify any additional actions, based on current analysis, that may need to be taken to fully implement the Expanded TO-BE architecture.

Doctrine and Policy

Leftwich et al. (2002) recommended several changes to doctrine and policy:

- Rewrite Air Force Doctrine Document 2 (AFDD 2), AFDD 2-4, and AFDD 2-8 to include basic objectives and functions of

CS execution and planning and control and organizational alignment.

- Increase emphasis on CS execution planning and execution role.
- Develop and write policy for CS execution planning and control.

To address these recommendations, the Air Force initiated a review of its doctrine and policy and then started revisions to reflect the TO-BE CSC2 operational architecture. At the time of this writing, changes were being implemented to AFDD 2-4, and it was planned that, as AFDD 2, AFDD 2-6, and AFDD 2-8 come up for revision, they would also include the CSC2 concepts. Further, Air Force policy and procedures were to be written or modified in AFI (instruction) and AFTTP (tactics, techniques, and procedures) format, to further detail the doctrinal concepts.[1]

Air Force leadership has set the direction for change by initiating modifications to doctrine and policy. However, there are additional measures the Air Force can take to align doctrine with the Expanded TO-BE operational architecture.

Currently, there is no standard process or format for operational planners to communicate operational parameters to CS planners (to feed beddown planning, TPFDD, munitions, spares, transportation).[2] Those planners projecting resource requirements and later planning the TPFDD often work from different assumptions and with information of varying fidelity regarding operational requirements. This deficiency hinders timely, accurate CS planning. Beyond the lack of standardization, one obstacle to a coordinated dissemination of operational requirements is the transfer of planning responsibilities from the A-3/A-4 to the AOC during the transition from deliberate to crisis action planning. Early in the crisis action planning process, as planning responsibilities transition, CS planners often do

---

[1] Kevin Sullivan, "Concept to Reality," *Air Force Journal of Logistics*, Vol. 27, No. 2, Summer 2003.

[2] Interview with PACAF/LG-ALOC personnel, May 2003; interview with USAFE/LGSF personnel, April 2003.

not know operational requirements with enough certainty to make CS plans. Sometimes it is simply difficult ("like pulling teeth"[3]) to get operations planners to supply *any* information that might commit them to plans they could change. Then, when operational parameters are supplied, they are often communicated inconsistently to CS planners who need them.[4]

Air Force Instruction 13-1AOC, *Operational Procedures—Aerospace Operations Center*, provides guidance for the operation of the AOC and clearly denotes the functions involved in operations C2. This document does not make explicit what information operations planners must provide to CS planners outside the AOC (e.g., CS planners in COMAFFOR) to drive timely and accurate CS planning.

Creating a framework, reinforced in doctrine, to delineate specifically what information operations planners provide, in what format and to whom, could address this shortfall. The OPT described in Chapter Three gives an example of this. Planners should proceed with caution here. Operators have an understandable reluctance to divulge operational plans that will probably change. Thus, operators and logisticians must agree with each other on how to proceed, so that operators will see the benefits of supplying more information (and the shortcomings in the current system in which logisticians either do nothing or proceed anyway with incomplete or conflicting information) and that logisticians will be flexible enough to appreciate the inherent uncertainties and adjust accordingly. Solidifying this linkage between operations and logistics in crisis action planning would

---

[3] Interview with PACAF/LG-ALOC personnel, May 2003.

[4] Apparently, during OIF planning, USAFE ammunitions planners were able to get good information ("90 percent") as early as September 2002 to make their plans (interview with USAFE/LGMW personnel, April 2003). Others were not. For example, the AFMC Logistics Support Office (LSO), which had to make transportation cost projections for deployment and sustainment, was forced to use best guesses based on little real information (interview with AFMC/LSO personnel on May 7, 2003). USAFE fuels planners also reported general problems getting operational requirements (Interview with USAFE/LGSF personnel, April 2003).

enable a step forward in the coordination, timeliness, and accuracy of CS planning.

Currently, the Air Force's capability assessments are done ad hoc for each contingency in each theater. While some customizing for COMAFFORs is necessary, CS planners often reinvent the wheel when it comes to capability assessments. The content and format of these should be designed rationally and codified. This will enable personnel to be trained consistently and to think and communicate in the same terms across nodes. The product library on the CD-ROM offers a starting point for this effort. The frequency with which these assessments are done should be analyzed and standardized, too.

The shortfalls and proposed solutions described above are summarized Table 4.1.

Organizations

Leftwich et al. (2002) made two broad recommendations to address organizational problems:

- Establish standing CS organizations with clear C2 responsibilities.
- Develop procedures for centralized management of CS support resources and capabilities.

Table 4.1
Doctrine and Policy Shortfalls and Proposed Solutions

Shortfall	Proposed Solution
CS planners do not get operations parameters early enough in the crisis action planning process	Develop and write policy to ensure a single set of operational parameters is passed to CS planners early in the planning process
Capability assessments are ad hoc for each command and contingency	Develop appropriate capability assessment and reporting policy

After reviewing the TO-BE operational architecture, the Air Force endorsed the proposed CSC2 nodal construct, agreed that the alignment of C2 responsibilities must be defined clearly and assigned to standard CS nodes, and made plans to designate specific organizations to fulfill the responsibilities of each node. Some actions had already been taken to move in this direction. While some organizational developments were inspired by the analysis, others have simply occurred in response to the demands of recent operations but correspond to concepts in the TO-BE operational architecture.

Headquarters PACAF recognized that progress toward a COMAFFOR operational concept required a CSC2 architecture capable of facilitating information flow across all CS functions and levels of command but that no Air Force–wide standard for a CSC2 operational information process flow or systems architecture existed. Headquarters PACAF agreed conceptually with the nodes and information flows of the TO-BE architecture but desired further validation and linkage to information systems that support CSC2. It requested in September 2003 that the Air Staff authorize and direct a six-month C2 Battlelab-sponsored proof-of-concept assessment of the TO-BE architecture for CS applicability for PACAF.[5] Headquarters AF/ILG supported PACAF's request to test the architecture, and ILGX established and chaired a working group for this effort.

Operational Support Center

OSCs are an example of a continuously developing organizational node that corresponds to TO-BE architectural concepts. Within several MAJCOMs, OSCs have evolved as a matter of necessity for handling the day-to-day operations that did not fall under their Title 10 "organize, train, and equip" responsibilities but were too great for a numbered air force (NAF) to manage alone.[6] ACC; United States Air Forces, Europe (USAFE); and PACAF each has its own OSC, called

---

[5] Headquarters PACAF official message transmitted July 8, 2003.

[6] Sullivan (2003).

an OSC, USAFE Theater Air Support Center (UTASC), and PACAF OSC (POSC), respectively, at various stages of evolution.

There are several examples of COMAFFOR and MAJCOM functions evolving in conformity to the TO-BE architecture. During OAF, the USAFE/LG (Logistics Group) staff was organized into control cells to manage the CS infrastructure. In the absence of guiding policy, these cells developed innovative reporting and control procedures to meet customer needs, which were critical to CS execution decisionmaking.[7] The cells resembled aspects of the CSC2 TO-BE operational architecture now being implemented. During OEF, COMAFFOR A-4 roles (supporting Central Command [CENTCOM]), although developed ad hoc, in many ways resembled those described in the TO-BE operational architecture. Also during OEF, the UTASC A-4 performed functions that the CSC2 operational architecture indicated could be assigned to an OSC.[8]

More recently, PACAF/LG proposed two cells to reside within the POSC. A movement cell (proposed by PACAF/LGRR) would track Pacific AOR unit line number taskings, scheduling, and movement in support of contingency operations, essentially closing the loop between PACAF wings and the POSC. This cell would also be able to express to operators what would be the operational impact of TPFDD closure delays. The movement cell was approved by PACAF/DO for implementation.[9]

CS planners in PACAF/LG-ALOC proposed a beddown assessment team concept. They cited two deficiencies in the TPFDD development process: general C2 and the need for feasibility assessments to determine whether a TPFDD would support planned operations. The beddown assessment team would have, among

---

[7] Department of the Air Force, Air Force Doctrine Document 2, *Organization and Employment of Aerospace Power*, September 28, 1998, pp. 32–35; Headquarters, USAF/CVAE staff, transcript of interview, January 4, 2000; Headquarters, USAFE/LGT staff, transcript of interview, February 1, 2000.

[8] Robert S. Tripp, Kristin F. Lynch, John G. Drew, and Edward W. Chan, *Supporting Expeditionary Aerospace Forces: Lessons from Operation Enduring Freedom*, Santa Monica, Calif.: RAND Corporation, MR-1819-AF, 2004.

[9] Email discussion with PACAF/LG-ALOC personnel, September 11, 2003.

others, responsibility for providing standardized beddown planning instructions, ensuring that all functional area managers use the same planning assumptions, resolving beddown limiting factors, and coordinating with NAFs, wings, and other services. The proposed team hopefully would aid in coordination when building TPFDDs, provide senior leadership with feedback on beddown supportability, and ensure that forces are correctly identified, flowed, and supported to meet warfighter requirements.[10] While this team is still conceptual, many of its processes are already followed by PACAF/LG-ALOC.

Other evolving organizations are the LSCs and the USAFE Theater Ammunition Control Point, which have provided some assessment capabilities and asset visibility (for spares and munitions, respectively) to the OSCs, essentially functioning as virtual nodes.[11]

The Air Expeditionary Force Center (AEFC) at Langley Air Force Base also performs some functions of an OSC. As a force nominator, the AEFC acts as a COMAFFOR Rear in TPFDD development. As a major information manager for personnel, it aids in (personnel) resource allocation. The AEFC can also provide reachback capability to the COMAFFOR Forward to answer personnel queries.

The Air Force's OSCs have transformed in response to real-world demands. During recent contingencies, they have served many important functions. However, our analysis revealed several remaining shortfalls in the development of Air Force OSCs. These shortfalls and proposed solutions, along with those for other organizational nodes, are summarized in Table 4.2.

The Air Force could benefit by standardizing many functions, recording these in doctrine and policy, and creating a set of metrics that applies to all OSCs. The Air Force should standardize OSC metrics, skills, and training to the degree possible, so that operators and logisticians know what skills are in OSCs and what capabilities commanders can reach back to get.

---

[10] Email discussion with PACAF/LG-ALOC personnel, September 11, 2003.

[11] Tripp et al. (2004).

Table 4.2
Organizational Shortfalls and Proposed Solutions

Shortfall	Proposed Solution
Visibility of manpower and equipment resources scattered	Bring oversight and management of manpower and equipment under one CONUS OSC
Spares POM process does not incorporate closed-loop feedback	Incorporate closed-loop feedback in spares CCP processes
CSC lacks adequate analytic capability	Add analytic capability to CSC

Currently, the AEFC is involved only in scheduling and planning personnel and in expeditionary combat support. If the AEFC were enlarged and given additional resources, it could maintain visibility of all personnel and worldwide equipment stocks, and the Air Force would have one point of contact for AEF manpower and equipment issues. With tasking authority, the AEFC could be a virtual node of the ACC OSC (or the Air Force Forces Command OSC, if the new organizational construct is approved) and accomplish all force nomination and tasking, manage AEF sustainment issues, and maintain unified visibility over unit equipment (i.e., non-WRM equipment not tasked to specific plans). This would enable a more established and capable reachback capability.

Commodity Control Point

CCPs already exist within different organizations that manage Air Force resources, and the Expanded TO-BE architecture proposes enhancements to them. AFMC exemplifies changes that can align CCPs with the architectural concepts (in response to day-to-day needs, the Spares Campaign, the Chief's Logistics Review, and the architecture itself). The following text briefly charts its evolving concepts.

Headquarters AFMC established the Logistics Support Office almost a decade ago. AFMC/LSO monitors shipment pipelines (to correct back-order problems, depot processing issues, and contract problems), tracks delivery times by both commercial and military transportation, and coordinates with the Air Mobility Command, commercial carriers, and personnel in the AOR to alleviate shipping problems. If a shipment is delayed, AFMC/LSO's CONUS Distribution Management Cell is empowered to reroute shipments. AFMC/LSO relays delivery time information to customers to help them make better decisions about transportation modes for future shipments. This process monitoring and get-well planning typifies TO-BE CCP duties.

In OEF, AFMC/LG assumed many of the responsibilities identified with a spares CCP in the TO-BE architecture such as tracking spares shipments end-to-end, forecasting demands, and working more closely with customers and suppliers.

AFMC also drew up a concept of operations for a spares CCP[12] (called a Virtual Inventory Control Point) with features like those outlined in the TO-BE operational architecture. In the CONOPS, spares management is aligned with weapon systems and managed by weapon system supply chain managers. Also, one WS-SCM position has been created and filled at each ALC[13] to start implementing some of these concepts.

To maintain an awareness of customer needs, AFMC, along with its customer MAJCOMs, also set up a High Impact Target list. Each MAJCOM identified a set of its most important repair parts for AFMC to monitor in the various ALCs. This program is popular with the customer MAJCOMs, and AFMC has automated many of the processes associated with maintaining the list and gathering status reports. This is another aspect of CCP performance monitoring that mirrors the TO-BE architecture.

Learning from past experiences, AFMC established the Warfighter Sustainment Division to specifically address problems with wartime CS. The Warfighter Sustainment Division, comprising an Operations Branch and a Logistics Analysis Branch, was created to be a single interface between AFMC and its customers. The Operations

---

[12] Headquarters, Air Force Materiel Command, Virtual Inventory Control Point concept of operations, draft, February 6, 2003.

[13] F-15 at Warner Robins, KC-135 at Oklahoma City, and F-16 at Ogden.

Branch tracks shipments, coordinates repairs among maintenance organizations, and responds to customer queries; the Logistics Analysis Branch provides forecasting and attempts to identify shortfalls and issues.

AFMC has implemented many important measures to transform its management of spares. Our analysis revealed several additional steps AFMC could take in transforming the spares CCP to align with the Expanded TO-BE operational architecture.

These successes aside, AFMC currently lacks the ability to inform the Air Force POM process with spares data to enable MAJCOMs to shape their spares POM to predictably achieve Air Force–stated WSA goals and training requirements. Whereas AFMC acquires and produces most CS goods and services used by the operating commands, the MAJCOMs develop the POMs and BESs that program and fund them. Although AFMC does not directly affect the levels of spares support it will procure or provide in the year of execution, once the total Air Force POM has been coordinated, AFMC influences the budget submission when it updates prices used by MAJCOMs in developing the BES. The drawback of this process is that it produces a budget only for what the MAJCOMs agree to spend, not what the total bill will be.[14] The MAJCOMs often do not understand or appreciate the contribution to readiness made by basic investments in pipelines, safety stocks, or spares not directly tied to their own operational goals. Additionally, according to the Programming and Financial Management Team of the Air Force Spares Campaign, the current process assumes incorrectly that a MAJCOM's total program covers all spares needs and that all budgeted funds will go to AFMC in sales. When AFMC tries to fund remaining spares in the BES, their efforts are viewed only as a price increase.[15]

---

[14] Camm and Lewis (2003).

[15] Ed Koenig et al., "Recommended Changes to Air Force Spares Programming, Budgeting, and Financial Processes," Programming and Financial Management Team, U.S. Air Force Spares Campaign, AF/ILS (Installations and Logistics), Washington, D.C., 2001.

To address the Herculean task of developing an integrated cost estimate of spares requirements, AFMC built a consensus process to pull together requirements from all involved parties. Because of the fragmentation of stakeholders and their differing views, spares requirements have been chronically underfunded since the advent of stock funding. Because of the decline in mission-capable rates and the persistent lack of resources, analyses of AFMC and MAJCOM processes were performed, and deficiencies were discovered in both. Subsequently, the SRRB integrated product team received direction to merge the processes, so that one budget representing coordinated requirements from MAJCOMs and AFMC could be developed. The resulting SRRB brings together the MAJCOMs, HQ-AF, and AFMC's ALCs to develop the total Air Force requirements for spares for the POM cycle, in hopes of obtaining "full funding" in the future. However, until they perceive a benefit to readiness or increased capability, the MAJCOMs are not likely to use their limited resources to invest in safety stocks or other Air Force–wide inventory they do not control.

Decisionmakers should know what desired levels of capability will cost, so that during budgeting and execution, when resources do become constrained, there is a firm foundation for restructuring and a documented track to the reasons for the reduced capability.[16] A cost estimate of spares requirements (using the data described in Chapter Three) with documented assumptions and clear explanations of how the estimates were compiled could build a strong, defensible base for submitting budgets and for revising budgets when fiscal constraints are imposed.

The spares CCP could make a difference in this process. Armed with actual outcome data gathered from its many sources, this CCP could provide an independent cost estimate, which would provide a

[16] Camm and Lewis (2003) make a strong argument for a wider, more inclusive vision of logistics. Instead of viewing weapon systems and CS as "tooth and tail," they propose building a framework that makes the connection between logistics resources and combat capability more visible ("without the tail, the tooth can't bite"). The goal of this strategy is to obtain both warfighters' and congressional support during the battle for resources in the PPBE.

baseline for establishing the availability goals and would serve as a test of reasonableness for the SRRB requirement. Actual usage data, combined with prices charged by suppliers and sources of repair, and trend data showing historical price behavior could be contrasted with the model outputs. Because the CCP will also be able to link spares requirements to availability of aircraft showing the effect of limiting factors and constraints, it may even be able to convince the MAJCOMs of the benefits of optimizing their inventory investment.[17]

Also, AFMC processes lack feedback mechanisms to track spares production and WSA performance and make midcourse corrections to meet predetermined WSA targets. This goes hand in hand with decision support tools. Once tools exist that can effectively link depot maintenance to operational metrics, processes may be altered to enable AFMC to adjust maintenance priorities to meet training or readiness goals.

Combat Support Center

At the Air Force level, the operational architecture calls for the CSC to monitor CS requests for a particular contingency and assess the impacts of those requests on the ability to support that and other contingencies. During OEF, the existing Air Force CSC assumed many responsibilities of the future CSC, such as integrating multitheater requirements, identifying global resource constraints by commodity, conducting integrated assessments (base support), and recommending allocation actions for critical resources.

The CSC performed these functions and intervened when necessary to allocate scarce resources to the AOR when those resources might have otherwise been designated to support other AORs and other potential contingencies. The CSC did the actual assessments for FOL support assets and relied on the supporting MAJCOMs to supply weapon system supportability assessments and assess the impact of OEF operations on peacetime training and other potential contingency operations. The operational architecture calls for the CSC to

[17] Tests of reasonableness (or crosschecks) for cost estimates are standard practice in weapon systems acquisition.

conduct these weapon system and FOL support assessments. The CSC conducted weapon system assessment functions at the Air Staff, and each MAJCOM conducted weapon system assessments, as well. In OEF, the CSC performed a worldwide assessment of FOL support capability and determined when the Air Force could provide support for other services—for example, for Army special operations forces—and made recommendations to the Joint Chiefs of Staff accordingly.

The Agile Combat Execution Support (ACES) team at the Air Staff performed the functions of the CSC for OIF. In addition to conducting integrated assessments, such as base support, the ACES team tracked and monitored many action items identified in the logistics supportability analysis and worked to find solutions for competing demands of scarce resources. The ACES team took a global view of CS and, while working for the AF/IL, was able to cross MAJCOMs and theaters to find optimal solutions.[18]

Our analysis revealed another remaining shortfall in Air Force CS organizations. Although the CSC was able to support OEF, it should be supplemented with analytic skills in the future. Adding a capabilities assessment function and a limited number of CS personnel with skills in quantitative methods to accomplish more refined capabilities assessments should enhance this capability. This same team could support quantitative assessments needed for the POM process during noncontingency operations.

Training and Education

Leftwich et al. (2002) made several recommendations to address training shortfalls:

- Develop CS course curriculum for C2.
- Expand the role of CS in wargames and exercises.
- Take advantage of joint services logistics wargames.

---

[18] Tripp (2004).

- Incorporate C2 gates in CS officer and enlisted career development.
- Develop C2 job performance aids for CS.

The Air Force has made much progress in making changes to training. The Air Force plans to take advantage of joint services logistics wargames (e.g., the Future Logistics Wargame) to evaluate new concepts and expand skills training in tactical-level exercises (e.g., Silver Flag). There will be an education working group as part of the implementation team to address the development and enhancement of formal education programs. The Air Force Advanced Maintenance and Munitions Officers School at Nellis Air Force Base, Nevada, has already implemented significant C2 instruction in its curriculum,[19] and the Air Force Academy has included ACS work as part of its "Logistics of Waging War" course.[20]

The Mission Support Group (MSG) commanders course and the new CS Executive Warrior Program will provide training for MSG commanders, potential expeditionary MSG commanders, and A-4s. Eagle Flag will provide the final field training exercise for CS personnel prior to their AEF rotation and give them the opportunity to test their ability to open and establish an airbase and provide initial C2. On the academic side, one of Air Command and Staff College's eight new specialized studies will provide an overview of ACS for officers and civilians within and outside the ACS community. The Air Force Institute of Technology is revamping short courses to be in line with the new combat wing organization and logistics processes. Finally, the Advanced Logistics Readiness officer course will provide a special logistics expertise to the warfighter.

Our analysis revealed one important remaining shortfall in Air Force CS training and education: the need to codify the concept of an OPT, in doctrine and policy. To reinforce this, we recommend that the Air Force ensure that operators are trained to create OPTs in

[19] Sullivan (2003).

[20] Tripp (1999).

a timely manner (understanding their uncertain planning environment) and to understand the expectations of CS planners that these documents will aid. Operators should understand what CS planners need and when, and CS planners should understand the limitations and uncertainties within which the operators must work. Only by training both groups to understand both sides of the planning equation and communicate effectively will this link between operational and CS planning be made and sustained.

Information Systems and Decision Support

Leftwich et al. (2002) made the following recommendations for an information system and decision support tools:

- Develop tools to provide required capabilities
  - Relate operational plans to CS requirements
  - Convert CS resource levels to operational capabilities
  - Conduct capability assessment and aggregate them on a theater or global scale
  - Conduct trade-off analyses of operational, support, and strategy options.
- Focus integration efforts on global implementation of a few selected tools.
- Standardize tools and information systems for consistent integration.

To implement this work, the Air Force has created an implementation team that has been patterned after the approach taken in the Chief's Logistics Review and the Spares Campaign. It will be this team's charge to take the operational architecture; solicit comments from Air Force component commands, Air Staff, and MAJCOMs; and integrate lessons learned from previous and ongoing operations to develop and refine an executable implementation plan. This plan will be time phased and will focus on specific objectives. There will be a road map with associated metrics to indicate current status and progress toward capability-based goals.[21]

The Air Force already had a number of tools that performed some of these recommended functions. And in the time since the release of the previous report, the Air Force has further developed some and begun development on others. The following are merely examples of these; certainly there are more in various stages of development too numerous to describe here.

Two tools exist that could link depot processes and constraints to AFMC's spares planning process. The Execution and Prioritization of Repair Support System (EXPRESS) Planning Module (EPM) was created at Ogden Air Logistics Center and has also been used at Warner Robins Air Logistics Center; it is being tested at Headquarters AFMC. This tool has the ability to predict customer needs, prioritize them, and evaluate depot resource availability. Advanced Planning and Scheduling (APS) is a commercial off-the-shelf tool with capabilities similar to EPM.[22] EPM is a homegrown information system and is tailored to some requirements specific to the Air Force, but personnel at Headquarters AFMC have concerns about whether APS, a commercial package, oriented toward commodities rather than weapon systems, has the ability to do the same. EPM is still officially a prototype information system being used at Ogden and Warner Robins. AFMC/LG has approved the merging of the EPM functionality into the Headquarters-approved standard configuration of EXPRESS (i.e., it will eventually be subsumed by EXPRESS, thereby becoming part of the standard suite of information systems). APS is being evaluated at Oklahoma City, with plans to test at other ALCs. Direction from OSD suggests that APS will be used.

The FMSE calculator is an Excel-based tool created at the Air Force Petroleum Office (AFPET) that translates operational parameters (e.g., aircraft type, sorties) into CS requirements (e.g., UTCs). It

---

[21] Sullivan (2003).

[22] APS has capabilities beyond EPM that are currently accomplished by other AFMC systems.

is continuously developing but has already been used for execution by fuels planners. This tool (though a "working prototype") exemplifies the kind of tools that will enable the TO-BE vision to become a reality.

Another information system capability important to the TO-BE architecture is one into which low-level CS personnel can input resource and process information that automatically generates capability assessments for higher-level visibility. Munitions reports and petroleum, oil, and lubricants (POL) reports already do this. PACAF has Web-based munitions and POL reports that are automatically generated in different formats for different audiences from lower-level resource information invisible to the final viewer. This is the kind of information system function necessary to enable TO-BE concepts.

Our analysis revealed several remaining shortfalls in Air Force CS information systems and decision support tools. These shortfalls and their proposed solutions are summarized in Table 4.3.

A closed-loop planning and control environment is important to the Expanded TO-BE architecture. EPM and APS, described above, are two information systems that may meet architectural requirements for closed-loop spares planning. Analysis should be done to see whether these will meet the Air Force's needs.

Table 4.3
Information System Shortfalls and Proposed Solutions

Shortfall	Proposed Solution
Much effort expended to perform capability assessments; usually manual, often reinventing the wheel	Common operating picture for capabilities assessments
Disconnect between budget planning and depot maintenance planning and execution	Closed-loop spares planning
Deployment planning processes and information systems slow to react and lack ability to plan and replan rapidly and explore multiple deployment options	Systems engineered planning tools

Currently, information about Air Force resource and process metrics is often organized by commodity or end item and located on disparate information systems. Creating a single information system accessible to a wide audience would enhance leadership visibility over these resources. Such an information system would need to have enough automation to translate lower-level process and resource data into aggregated metrics and even some operational metrics (e.g., WSA, sorties). This information system would inform commodity managers, planners, and senior leaders who already have to make decisions in the AEF operational environment.

Rapid deployment and capabilities-based planning have been emphasized in our description of the TO-BE architecture. Having a rapid, flexible deployment process suggests the need for a rapid TPFDD planning tool. Adopting a planning strategy based on a portfolio of capabilities[23] suggests the need to develop a means to calculate the manpower and equipment required to generate each of the capabilities in that portfolio, to source and time phase each, and then to assess the transportation feasibility of each plan—all in a very short time frame. Existing information systems are unable to support these capabilities, although pieces of such an information system do exist within and outside the Air Force. A suite of tools that could automate as much of this planning work as possible would greatly expedite the deliberate and crisis action planning processes and hence usher along the transition to a capabilities-based, expeditionary Air Force.

RAND analysis has recently produced a prototype of a requirements TPFDD generator.[24] A fully developed version of this tool could enable the kind of quick planning prescribed by the Expanded TO-BE operational architecture. Also, the Joint Flow and Analysis System for Transportation (JFAST), developed under and managed by USTRANSCOM, is an automated information system that tests

---

[23] Rumsfeld (2001c).

[24] Don Snyder and Patrick Mills, *A Methodology for Determining Air Force Deployment Requirements*, Santa Monica, Calif.: RAND Corporation, MG-176-AF, 2004.

the transportation feasibility of executable TPFDDs. These both could supply pieces of this deployment planning architecture.

Future Logistics Enterprise

The Air Force plans for the CSC2 implementation effort to be fully integrated with the Future Logistics Enterprise and other CS enterprise architectures. They plan to develop systems and technical architecture views that are Enterprise Architecture Initiative compliant. Within the systems architecture will reside the CSC2 tools that provide responsive capability analysis, decision support for the resource arbitration process, CS execution feedback (equivalent of battle damage assessment for operators), and forward-looking assessments. These tools should strengthen communication channels between supporting and supported functions. AF/ILI and AF/ILGX will work together to integrate CSC2 architectures and the Future Logistics Enterprise to build the foundation for making CS truly agile.[25]

---

[25] Sullivan (2003).

CHAPTER FIVE

Summary and Conclusions

CSC2 is important in today's operating environment of capabilities-based planning, constrained resources, and frequent deployments. CSC2 is an enabler across the spectrum of operational concepts and phases of operation. As its CSC2 continues to improve, the Air Force will plan and execute operations more effectively and efficiently.

The Air Force has made changes to its CSC2 architecture over the last few years, moving toward concepts outlined in our previous work. Initiating organizational and doctrinal changes were the first steps in this direction, yet several steps remain.

The Air Force has plans in place to continue changing doctrine and policy. Incorporating into doctrine and policy instruction as to what operational parameters operators ought to communicate to CS planners during crisis action planning will enable more accurate, coordinated planning in all CS areas. Rationally creating and standardizing CS resource capability assessments in content, format, and timing will help create a powerful web of visibility over Air Force capabilities that will lead to more sensible, effective operations planning and CS resource configuration.

Many organizational nodes in the Air Force already function like nodes in the TO-BE architecture. Bringing oversight of manpower and equipment to a single OSC would enable efficient deployment planning. Incorporating more sophisticated, longer-term demand prediction into spares management would make it a powerful contributor to the POM and crisis action planning processes; incorporating closed-loop planning and execution capabilities would create a spares management system that could more reliably and precisely support warfighter needs. Adding analytic capabilities into the CSC would enhance its ability to assess global Air Force capabilities.

Already developed and embryonic information systems and decision support tools exist within and outside the Air Force to close gaps between the AS-IS and TO-BE architecture. Harnessing these information systems to incorporate closed-loop planning into spares management would support a spares management system that could more reliably and precisely support warfighter needs. Engineering a system of decision support tools for rapid deployment planning and assessment would speed the transition to a capabilities-based, expeditionary Air Force.

Such a transition will require a concerted effort on the part of leadership to reinforce and push changes to doctrine, policy, and process. While changes to better integrate operators and logisticians will be challenging—their patterns and processes are often not simply procedural but may be ingrained in culture—the integration of these two is important.

Our approach and recommendations emphasize high-quality data and analysis, frequent, accurate capability assessments, and agility in planning and reconfiguring CS assets. While we have sought here to expand on more general concepts, only the Air Force has the ability to continue fleshing out changes at the lowest, detailed levels and to make appropriate changes.

APPENDIX A

Lessons and Observations from Recent Contingencies

This appendix traces the performance of the combat support command and control (CSC2) operational architecture as it has evolved during three recent contingencies—operations Allied Force (OAF), Enduring Freedom (OEF), and Iraqi Freedom (OIF). We start with OAF, whose observations are summarized in Table A.1.

Operation Allied Force

The transition to wartime combat support (CS) operations in OAF was difficult, partly because of a departure from doctrine that shifted command from the numbered air force (NAF) to the major command (MAJCOM) during operations. Shifting organizational responsibilities during conflict created problems, including attempting to use organizational relationships that did not exist on a day-to-day basis, delays in developing communications paths, and using people who may not have been trained for wartime jobs. Such problems may have arisen even if the NAF had supported OAF because of the staff augmentation necessary to make the current doctrinal organization effective.[1] Because the Air Force has been at a high operations tempo for the last 12 years, and this set of circumstances is likely to persist into the foreseeable future, there is a need for standing (permanent)

[1] Lawrence M. Hanser, Maren Leed, and Charles Robert Roll, *The Warfighting Capacity of Air Combat Command's Numbered Air Forces*, Santa Monica, Calif.: RAND Corporation, DB-297-AF, 2000.

CS organizations to provide operational continuity and seamless transition through the spectrum of military operations. The transition was also hindered by confusion over responsibility for theater distribution management, leading to an ineffective theater distribution system.[2] Rapid configuration of both theater and global distribution systems is essential to meet Air and Space Expeditionary Force (AEF) operational deployment and employment goals, further highlighting the need for these responsibilities to be clearly delineated between the services or for the Air Force to maintain the skills to develop and configure such a system.

In OAF, the limited ability and opportunity for interaction between the Air Force CS and operations planners led to plans being developed with minimal CS input, resulting in excessive revamping and slow progress.[3] Limited communications links between opera-

Table A.1
CSC2 Requirements Revealed by Lessons from Operation Allied Force

Lesson/Observation	CSC2 Requirements
Slow and difficult transition from peacetime to wartime operations	Identify permanent organizations that will perform critical CS tasks continuously during peace and war
	Expand Air Force involvement in theater distribution system planning and execution
Poor interface between operations and CSC2	Include CS input in initial planning
	Translate CS information into operational capability
Inability to react quickly to changes in the operational plans	Provide real-time visibility of theater and global resources
	Rapidly reconfigure CS infrastructure
Insufficient and inadequate CSC2 policy and procedures; systems; training; and education	Develop and formalize doctrine and policy; systems; and training programs

[2] Amatzia Feinberg et al., *Supporting Expeditionary Aerospace Forces: Lessons from the Air War Over Serbia*, Santa Monica, Calif.: RAND Corporation, 2002. Government publication; not releasable to the general public.

[3] Feinberg et al. (2002).

tions planners and CS planners hindered interaction. The CS person responsible for interaction in the operational planning group did not have a full depth of CS experience, information system links, or decision support tools to help facilitate interaction. This lack highlights the importance of formalized procedures for including CS factors in operational planning and execution and relating CS capabilities in operational terms.

Some CS organizations were slow to react to changes in the air campaign.[4] They were slow to reconfigure the CS infrastructure (to redirect materiel, adjust maintenance priorities, and alter distribution routes and modes) partly because personnel were inexperienced in these wartime functions. According to doctrine, the NAF was to exercise these responsibilities. MAJCOM personnel taking on these functions were not trained in many of them. Also, the Air Force took on some functions, such as planning the theater distribution system, that it may not have trained personnel for, assuming that the joint command would have the wartime responsibility.

Finally, OAF demonstrated that existing policies, procedures, and information systems were inadequate and that education and training were insufficient.[5] While good people compensated for these shortfalls, the deficiencies resulted in additional time spent to determine what should be done. We emphasize the importance of defining the role of CSC2 and of incorporating those activities into training and education.

The Air Force learned some lessons from OAF and began to change elements in the AS-IS architecture to better meet operational requirements. Even before the release of the TO-BE architecture, this evolving architecture was then put to use during OEF. The major lessons and successes and corresponding CSC2 requirements for OEF are summarized in Table A.2.

---

[4] Feinberg et al. (2002).

[5] Feinberg et al. (2002).

Table A.2
CSC2 Requirements Revealed by Lessons from Operation Enduring Freedom

Lesson/Observation	CSC2 Requirements
Slow and difficult transition from peacetime to wartime operations	Identify permanent organizations that will perform critical CS tasks continuously during peace and war
Confusion of responsibilities; duplication of effort	Identify permanent organizations that will perform critical CS tasks continuously during peace and war
CS resources stretched thin	Include CS input in initial planning
Incomplete information feedback loops	Complete CS and operations feedback loops
CSC2 system exhibited flexibility and responsiveness in CIRF operations	Continue using and standardizing CIRF operations Improve CONUS support location (CSL) linkages

Operation Enduring Freedom

OEF was a war entirely different from OAF in mission and scope. It shared some of the same CSC2 problems, though. Here, we describe some of the problems with command structure, resource planning and visibility, and process feedback, as well as progress made with intermediate repair facilities.

As in OAF and Operation Desert Storm, the OEF CS organizational command structure differed from the structure delineated in doctrine. This led to several difficulties as organizations developed ad hoc roles and responsibilities. Some organizations were not prepared for these evolving responsibilities. The command structure was further complicated by the global nature of OEF and other ongoing operations. Responsibilities were distributed across commands and regions, increasing information sharing burdens. In addition, multiple commitments to ongoing operations such as homeland defense (Operation Noble Eagle, a Northern Command [formerly Air Combat Command] responsibility), Central Command (CENTCOM) exercise Bright Star, support of Bosnia (a United States Air Forces, Europe [USAFE] responsibility), Operation Northern Watch (ONW)/Operation Southern Watch (OSW) (ONW was a USAFE responsibility; OSW, a Central Air Forces [CENTAF] responsibility), and others increased the necessity to prioritize among competing demands for time and resources. Despite all of this, the ad hoc Commander of Air Force Forces (COMAFFOR) A-4 command relationships closely resembled the CSC2 TO-BE architecture.[6]

In OEF, planners performed operational planning with little visibility of their resource impacts, which resulted in overextending many key CS resource areas. The Air Force supported many beddown locations, and several expeditionary CS areas became stressed. For example, personnel in career fields such as force protection, civil engineering, combat communications, and fuels were "borrowed" from future AEF buckets, causing personnel deployment rules (e.g., only 90 days of deployment time per year for each airman) to be violated.[7] This problem was not exposed in OAF, because fewer forward operating locations (FOLs) were opened, and fewer of these were bare bases (which require more manpower from the functional areas listed above). This highlights the need for a capability to determine rapidly the logistical implications of operational plans to integrate CS information into the operational planning process.

CS processes in OEF lacked adequate feedback mechanisms and were not linked rationally to operational goals. CS processes lacked data tracking capability to tie actual to planned performance levels to achieve specific levels of operational capability. Metric goals were based on history, not operational requirements. Support decisions (capacity, manpower, thresholds) were made not considering operational needs or requirements. Effective use of information feedback in CS planning and control is dependent on two things: (1) reliable access to information and (2) a framework for measuring CS process

---

[6] James Leftwich, Robert Tripp, Amanda Geller, Patrick Mills, Tom LaTourrette, and C. Robert Roll, Jr., *Supporting Expeditionary Aerospace Forces: An Operational Architecture for Combat Support Execution Planning and Control*, Santa Monica, Calif.: RAND Corporation, MR-1536-AF, 2002.

[7] Colonel Bruce R. Barthold, "Major Issues from the AFCESA/AFIT Sponsored Operation Enduring Freedom RED HORSE and Prime BEEF Lessons Learned Conference, 13–15 Nov 02," memorandum, December 2002, p. 19.

performances against "goals" or standards needed to achieve operational goals in the specific contingency operation.[8]

Aircraft maintenance forward support locations, referred to as CIRFs (Centralized Intermediate Repair Facilities),[9] were used successfully during both OAF and OEF. OAF showed that preselection and resourcing of centralized support facilities can improve flexibility and reduce deployment footprint. During OEF, CIRFs satisfied intermediate maintenance requirements for a number of reparable items for deployed fighter units with a reduction in forward deployed footprint and also supported forward bombers' phased maintenance. Goals were established linking warfighter needs to the CIRF maintenance process and theater distribution system performance. Developing similar links between CONUS support locations (e.g., air logistics centers [ALCs]), CS process performance and operational goals would enhance the supply system's efficiency and responsiveness.[10]

Operation Iraqi Freedom

By the time planning began for OIF, the changes in the Air Force CS architecture had resulted in noticeable improvements in CS support to operational planning. The major lessons and successes and corresponding CSC2 requirements for OIF are summarized in Table A.3.

The transition to wartime CS operations in OIF was less difficult than in other recent contingencies, for several reasons. First,

---

[8] Robert S. Tripp, Kristin F. Lynch, John G. Drew, and Edward W. Chan, *Supporting Expeditionary Aerospace Forces: Lessons from Operation Enduring Freedom*, Santa Monica, Calif.: RAND Corporation, MR-1819-AF, 2004.

[9] Some Air Force weapon systems (e.g., jet engines, avionics) were designed and manufactured with three echelons of maintenance: flight line, back shop or intermediate, and depot. Normally, the intermediate level of maintenance is forward deployed with flying units. The CIRF concept centralizes the intermediate-level maintenance operations of several squadrons and often locates this centralized capability somewhat rearward from the actual flying units, therefore decreasing the forward-deployed footprint.

[10] Tripp et al. (2004).

Table A.3
CSC2 Requirements Revealed by Lessons from Operation Iraqi Freedom

Lesson/Observation	CSC2 Requirements
Eased transition to war	Initial adoption of CSC2 architecture Better understanding of organizational responsibilities
CS input in initial planning	Adequate time before combat operations Good logistics supportability analysis
CSC2 system exhibited flexibility and responsiveness in CIRF operations	Continue using and standardizing CIRF operations Improve CSL linkages
Deployment planning system slow to react to changes in operational plans	Improve deployment planning process and systems

AF/IL had begun to implement the TO-BE CSC2 operational architecture. (At the time of OEF, this work was just being completed, and the recommendations had been viewed only by relatively few members of the Air Force senior leadership.) The implementation of this operational architecture contributed to the well-established, well-defined CSC2 relationships during OIF. The C2 structure developed during OIF closely resembled the CSC2 architecture. This was a significant improvement from the ad hoc command structures developed during OAF and OEF. Second, there was a continuity of people and organizations moving from OEF to OIF. Many standing organizations used during OEF were still in place. The leadership had recent combat experience; most leaders were in place for at least part of OEF, and many had also held key positions during OAF. Third, the command structure was well defined. During OIF, roles and responsibilities were established early, and organizations were given the authority necessary to perform their assigned responsibility. An essential factor that further enabled a smooth transition was the length of time Air Force planners had to plan. CENTCOM planners began to create an initial operational plan and define requirements in July 2002; logistics planners conducted a logistics supportability analysis (LSA), hosted by AFMC and facilitated by AF/IL, in August 2002; and combat operations did not start until March 2003.

There was more CS input into initial OIF planning compared with other recent operations. This was first enabled by the length of time planners had (timeline described above). Second, the LSA helped identify potentially limiting resources before operations were initiated and direct plans to mitigate these limiting resources, e.g., using ships to deploy munitions vice airlift. The LSA was used to help build the CS plan for OIF by allowing numerous participants from various MAJCOMs and functional areas to review the initial operational plan and identify actions that needed to be taken to support the plan. Following the LSA, the CENTAF A-Staff began working with the COMAFFOR to define alternative courses of actions (COAs) when CS shortfalls would impede the operational plan. Numerous actions were taken to move munitions, vehicles, bare base assets, and fuels support equipment forward.

CSC2 organizations exhibited flexibility and responsiveness during OIF via CIRF operations. CIRFs supporting OIF were located in USAFE. The ease with which the CIRFs operated should be attributed to having a well-thought-out concept of operations and an executable plan.

The deployment planning process and information system was slow to react to changes in operational plans during OIF. At the time of execution, the combatant commander changed the agreed-on flow of forces (1003V TPFDD [Time-Phased Force and Deployment Data]). The deployment planning process and supporting information system (which requires much manual intervention and coordination) was not designed for execution-level tailoring (especially given the Air Force's numerous unit type codes). Processes and information systems that support TPFDD development in crisis action planning and execution are essential to meet AEF deployment and employment goals.

APPENDIX B

Illustrative Examples of CSC2 Operational Architecture

In this appendix, we describe two notional examples of the application of the Expanded TO-BE CSC2 operational architecture. We use these examples to illustrate how planning and execution processes would work were the Expanded TO-BE architectural concepts implemented—specifically how organizations would interact, what information they would exchange, and what decisions this information would drive. We begin with deliberate planning.

An Example of Deliberate Planning

In the deliberate planning process, Air Force operators might consider many operational concepts (e.g., the United States defends an East Asian ally, attacks terrorists in a rogue state in the Middle East, or performs a humanitarian relief operation in Eastern Africa) for further exploration. They would develop each scenario to outline potential beddown schemes, weapon systems, type of munitions to use, and sortie rates for different phases of each operation.

Operations Support Center (OSC) planners could be working on plans for an antiterrorist operation in the Middle East where many bare bases must be opened. Using their visibility of AEF manpower capabilities, planners in the OSC would calculate, for each function, how current and projected AEF capabilities compare to the scenario requirements. In this case, as shown in Figure B.1 (p. 71), the total requirement for civil engineering explosives ordnance disposal personnel slightly exceeds the current AEF capability. Operations planners, after reviewing the analysis, might decide to consolidate forces at fewer FOLs, resulting in a less demanding scenario. In parallel, personnel at the Fuels Mobility Support Equipment (FMSE) Commodity Control Point (CCP) would check the ability of war reserve materiel (WRM) stores to meet contingency requirements. As shown in Figure B.2, they find that WRM stores are capable of meeting projected demands. Planners could decide to see how well remaining WRM would meet global requirements. The spares CCP, using mission design series (MDS) planning factors and expected sortie rates, would project spares requirements and develop graphics, for each MDS, like those depicted in Figure B.3 (p. 72). By correlating dollars spent (on spares) and weapon system availability, this shows that, while current manning and equipment levels would provide for an A-10 availability of 83 percent, this contingency requires an availability of 91 percent. Therefore, the CCP projects that the ALC supporting A-10s must surge its operations and would instruct it to draw up a COA accordingly.

After this and other plans have been reviewed and global requirements assembled, the OSCs and CCPs would then assess theater and global capabilities. The OSC would review beddown locations, depicted in Figure B.4 (p. 72). Planners might find that FOL capabilities are adequate to meet projected requirements. However, several CCPs could review their capabilities and find that get-well actions must be taken to meet these force requirements. The FMSE CCP, after reviewing global requirements, might find capabilities as illustrated in Figure B.5 (p. 73). It sees that its projected WRM stores would be incapable of satisfying combatant commander requirements. Planners could propose to incorporate an additional buy of several FMSE pieces in the next program objective memorandum submission, using those data as support. The basic expeditionary airfield resources (BEAR) CCP might find that the Air Force has BEAR packages adequate for contingency requirements but that they are malpositioned to give optimal deployment timelines. Figure

Figure B.1
Required and Available AEF Capability for Explosives Ordnance Disposal Personnel for Middle East Scenario

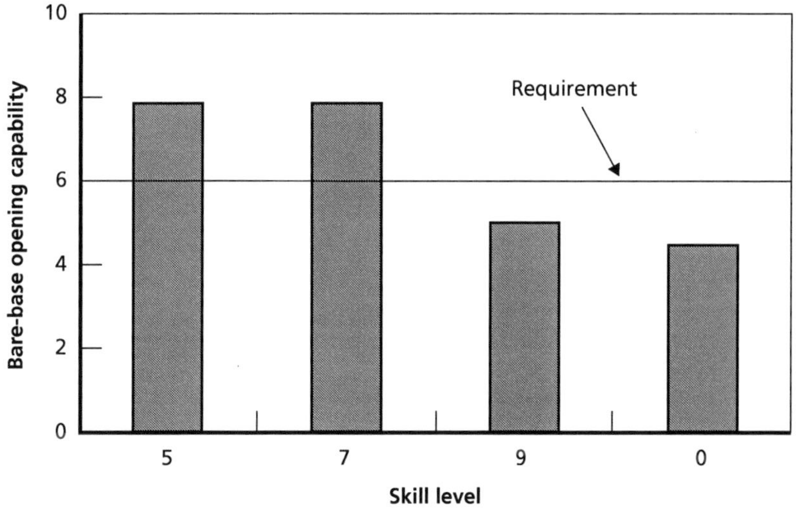

RAND *MG316-B.1*

Figure B.2
Required and Available AEF Capability for FMSE WRM for Middle East Scenario

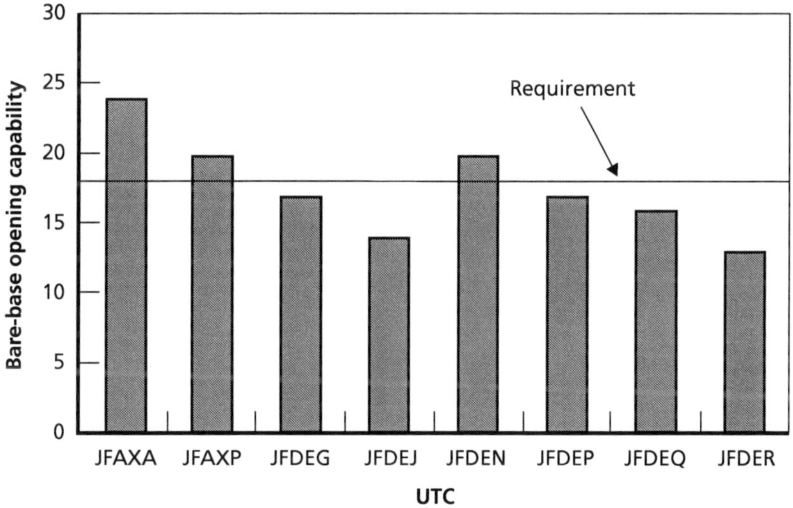

RAND *MG316-B.2*

Figure B.3
Middle East Scenario A-10 Spares Trade-Off

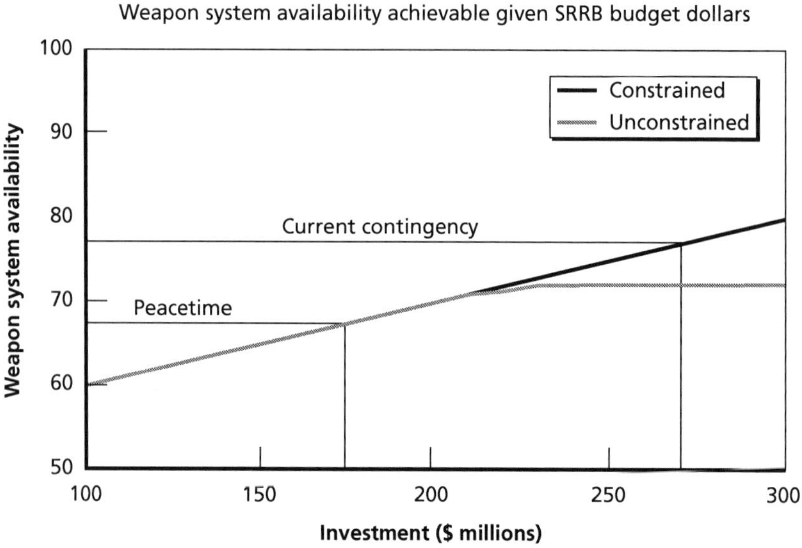

RAND *MG316-B.3*

Figure B.4
Theater Beddown Locations

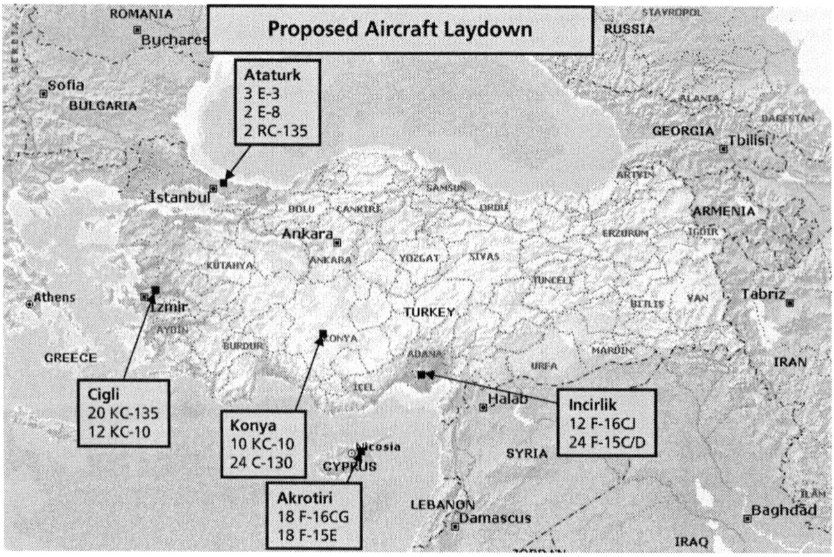

Figure B.5
Global FMSE WRM Requirement

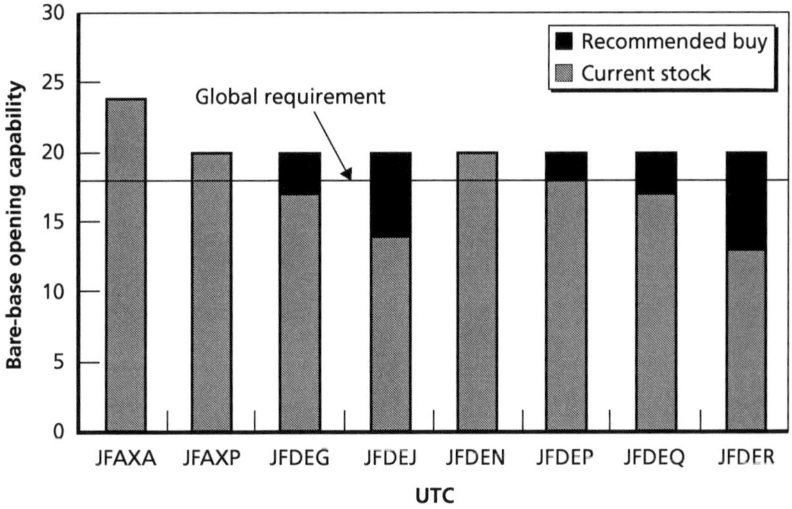

RAND *MG316-B.5*

B.6 depicts its current and future better positioning of its assets, which take advantage of travel times, throughput constraints, and political considerations.

These examples illustrate a few of the ways the OSC and CCP feasibility and capability assessments interact with the future planning Air Force planning process.

An Example of Contingency Planning and Execution

We provide another notional example to help illustrate the crisis action planning, deployment, employment, and sustainment phases of operation.

Suppose that because of the war on terror, the Air Force is already conducting antiterror operations in several countries, mainly supplying base operating support at many beddown locations, refueling operations, and precision guided bombing operations. If a

Figure B.6
Global BEAR Asset Inventories and Locations

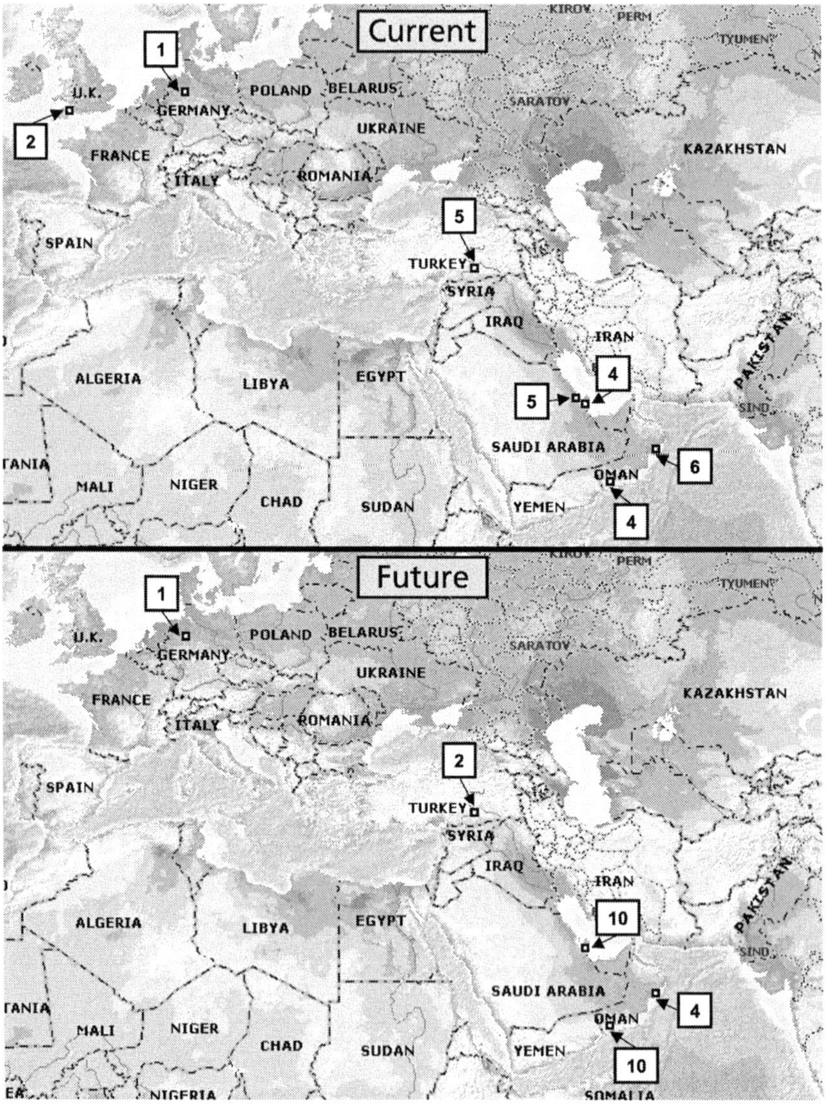

crisis with a particular country arose, it might require immediate action to mitigate risks of a nuclear attack on its neighbor. This would require large amounts of forces from the Air Force, so careful planning must be done, albeit quickly, to avoid compromising the Air Force's current and future warfighting capabilities.

Operations planners would quickly create several COAs for the combatant commander. These planners would access up-to-date beddown capability assessments regularly prepared by the OSC. Examples of this are in Figure B.7. (The white, medium gray, and dark gray represent green, yellow, and red from a "stoplight" chart.) They could quickly see which bases have which capabilities so they can propose realistic beddowns. The CSC would pass these requirements down to the CCPs, who would quickly perform feasibility assessments. The spares CCP, for example, could run the scenario and, for F-15C/Ds, produce output data like those in Figure B.8 illustrating trade-offs for weapon system availability supported by spares. This shows that, without surging ALC operations (requiring a reallocation of money), this weapon system may experience degraded sortie generating capability during the campaign, as depicted in Figure B.9.

Figure B.7
Beddown Capability Assessments

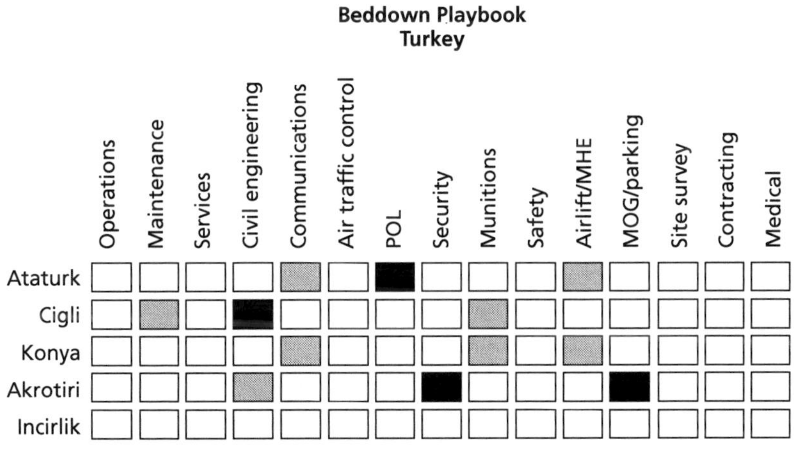

RAND *MG316-B.7*

Figure B.8
F-15C/D Spares Trade-Off

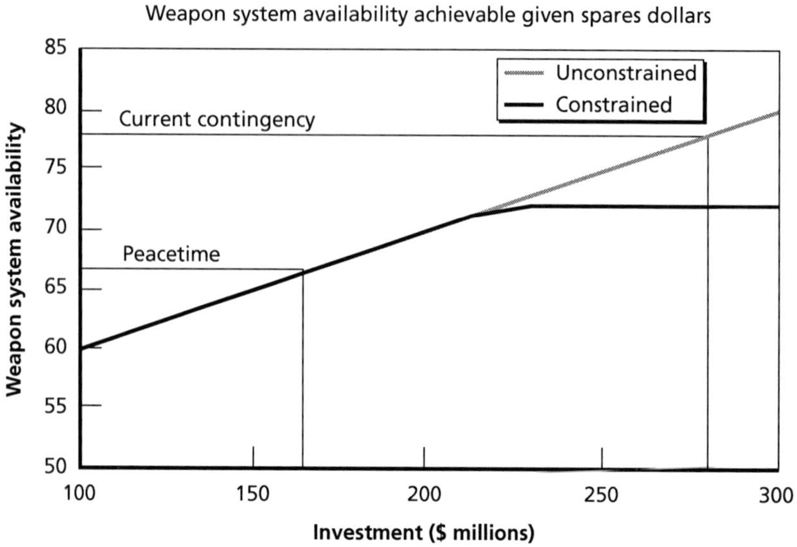

RAND *MG316-B.8*

Figure B.9
F-15C/D Degraded Sortie Capability

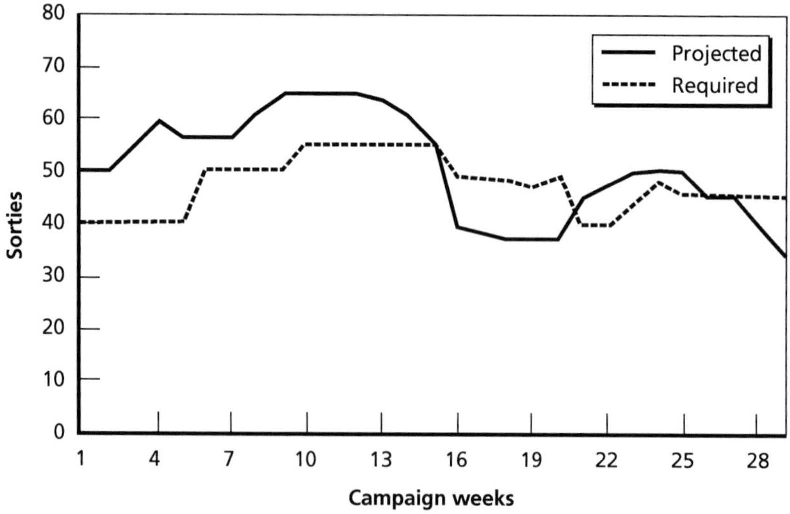

RAND *MG316-B.9*

Once units started to deploy, the OSC would monitor reception progress and base readiness at each FOL by receiving updates from each base. The OSC would feed these updates, like those depicted in Figure B.10, up to the COMAFFOR and combatant commander. The OSC would constantly manage the deployment process to ensure that units get to the right places at the right time to give the combatant commander the warfighting capability he needs.

During the planning stages, the OSC might determine that the CIRF that provides jet engine intermediate maintenance support to F-15s and F-16s in the ongoing antiterror operations could also support units deploying for the contingency scenario. The capability analysis, shown in Figure B.11, demonstrates that, although the CIRFs can support the other operations logistically, they would need both additional manpower and equipment to meet the greatly increased demand.

During operations, the munitions CCP, who is monitoring munitions stocks and expenditures at each FOL, might notice that

Figure B.10
Force Reception Progress

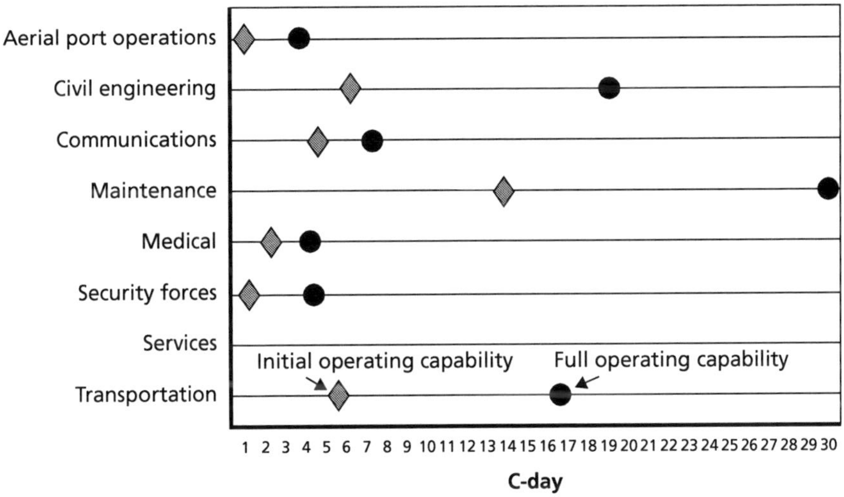

RAND *MG316-B.10*

Figure B.11
CIRF Capability

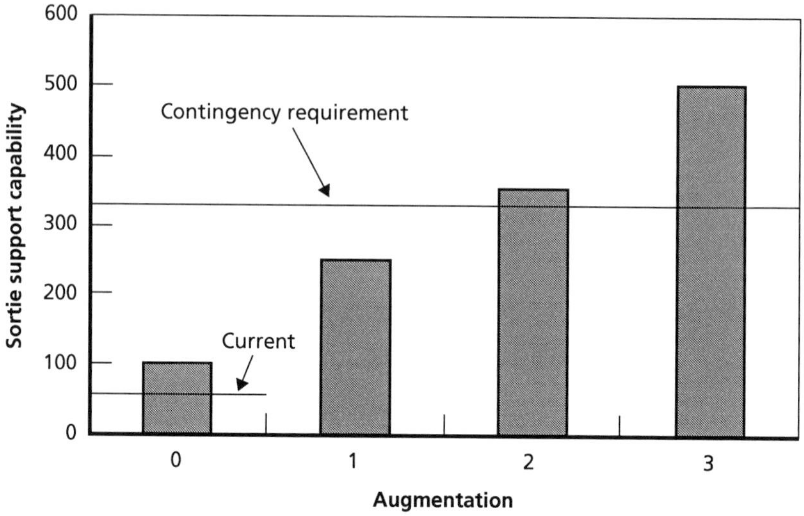

RAND *MG316-B.11*

Joint Direct Attack Munitions (JDAMs) are being expended faster than anticipated at several locations, because more targets are available than initially anticipated. The munitions analysis, shown in Figure B.12, predicts that sorties will be lost as a result of low JDAM stocks by day 25 of the campaign. The CCPs could propose either to swing stocks from a prepositioned ship or to use a different munition (e.g., laser-guided bombs, weather permitting).

During operations, if one of the transportation hubs the Air Force had ramped up for the contingency operation were attacked with a chemical munition, this would preclude the use of the entire hub (and the C-17s and C-130s bedded down there) until it could be decontaminated. This would cause all cargo there to be lost and incoming cargo in the region to be rerouted to another hub. The flow of sustainment might quickly overwhelm the second hub, causing a backlog. The OSCs would be monitoring the cargo backlogs and customer wait times (CWTs) to FOLs in the regions. They might observe behavior like that depicted in Figure B.13, which shows the

Figure B.12
Munitions Sortie Capability

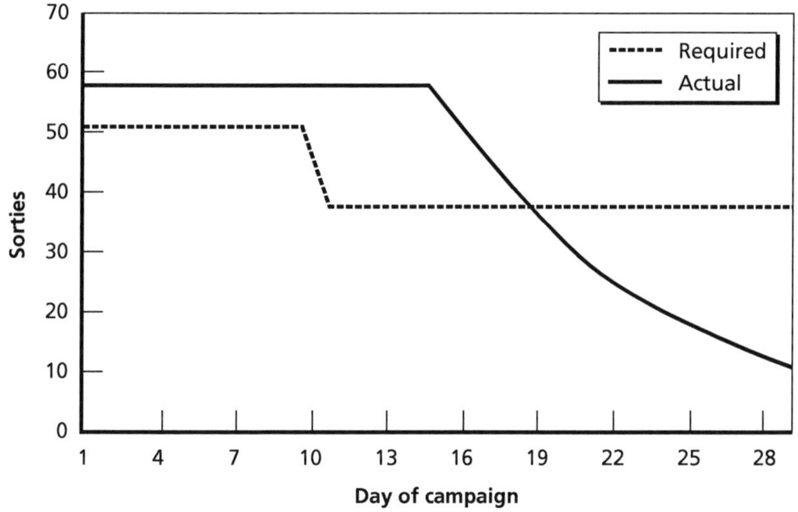

RAND *MG316-B.12*

Figure B.13
Cargo Backlogs/Customer Wait Time

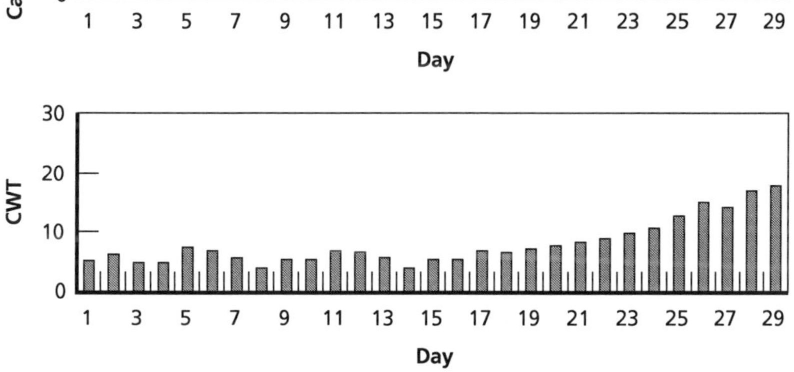

RAND *MG316-B.13*

backlog increasing and CWTs increasing. Spares CCP analysis, shown in Figure B.14, supported by the OSC performance monitoring, could predict that if spares removals occur at the same rate, but resupply times do not improve, sorties would be lost because of inadequate F-15 avionics by day 25 of the campaign. This would prompt get-well planning by weapon system supply chain managers. Finally, the OSC, as it received functional updates from each CCP, would integrate sortie capabilities into displays like that of Figure B.15, comparing capabilities of spares, fuel, and munitions. This would help both CS planners focus get-well planning to balance resources and operators project capabilities for further operational planning.

This example has shown some of the organizational nodes, information flows, and information products in the employment and sustainment phases of operation.

Figure B.14
F-15 Sorties Lost Due to Spares Deficiency

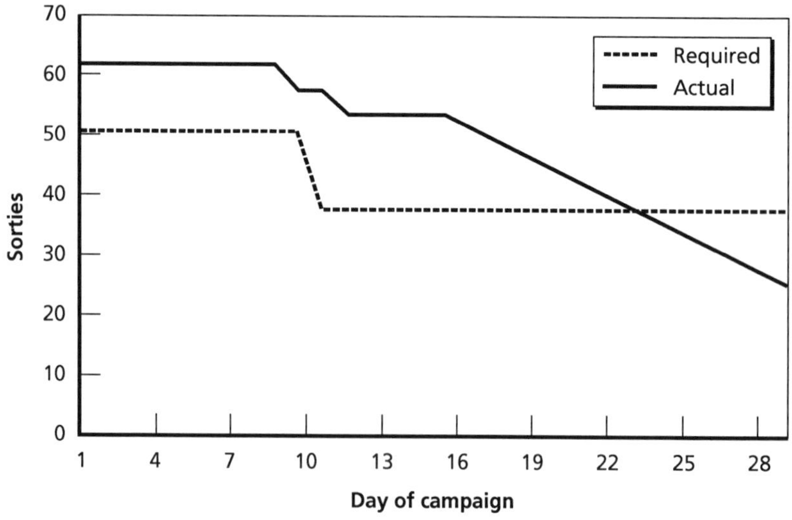

RAND *MG316-B.14*

Figure B.15
Integrated F-15C/D Sortie Capability Assessment

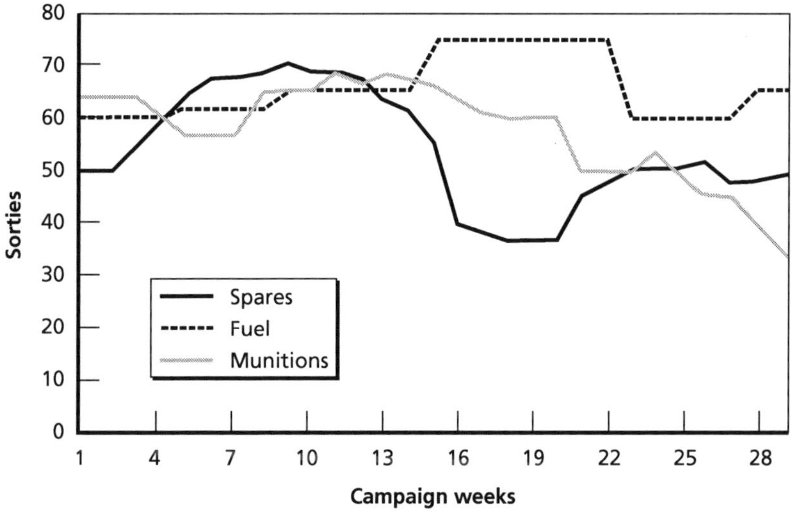

RAND *MG316-B.15*

Bibliography

Amouzegar, Mahyar A., Lionel R. Galway, and Amanda Geller, *Supporting Expeditionary Aerospace Forces: Alternatives for Jet Engine Intermediate Maintenance*, Santa Monica, Calif.: RAND Corporation, MR-1431-AF, 2002.

Barthold, Colonel Bruce R., "Major Issues from the AFCESA/AFIT Sponsored Operation Enduring Freedom RED HORSE and Prime BEEF Lessons Learned Conference, 13–15 Nov 02," memorandum, December 2002.

Camm, Frank, and Leslie Lewis, *Effective Treatment of Logistics Resource Issues in the Air Force Planning, Programming, and Budgeting System (PPBS) Process*, Santa Monica, Calif.: RAND Corporation, MR-1611-AF, 2003.

Davis, Paul K., *Analytic Architecture for Capabilities-Based Planning, Mission-System Analysis, and Transformation*, Santa Monica, Calif.: RAND Corporation, MR-1513-OSD, 2002.

Department of the Air Force, Air Force Doctrine Document 1, *Air Force Basic Doctrine*, September 1, 1997.

———, Air Force Manual 10-401, *Operation Plan and Concept Plan Development and Implementation*, Vol. 1, May 1, 1998a.

———, Air Force Doctrine Document 2, *Organization and Employment of Aerospace Power*, September 28, 1998b.

———, Air Force Doctrine Document 2-6, *Air Mobility Operations*, June 25, 1999.

———, Air Force Doctrine Document 2-8, *Command and Control*, February 16, 2001.

———, Air Force Instruction 13-1AOC, *Ground Environment Training: Aerospace Operations Center,* November 1, 2002.

Department of Defense, *C4ISR Framework Document Version 2.0*, December 18, 1997.

Elliott, Scott, "Force Modules Give Commanders 'Playbook,'" Air Force print news, January 24, 2003. Online at www.af.mil/news/Jan2003/12403907print.shtml (as of July 2005).

Feinberg, Amatzia, et al., *Supporting Expeditionary Aerospace Forces: Lessons from the Air War Over Serbia*, Santa Monica, Calif.: RAND Corporation, 2002. Government publication; not releasable to the general public.

Feinberg, Amatzia, Hyman L. Shulman, Louis W. Miller, and Robert S. Tripp, *Supporting Expeditionary Aerospace Forces: Expanded Analysis of LANTIRN Options*, Santa Monica, Calif.: RAND Corporation, MR-1225-AF, 2001.

Galway, Lionel A., Mahyar A. Amouzegar, Richard J. Hillestad, and Don Snyder, *Reconfiguring Footprint to Speed Expeditionary Aerospace Forces Deployment*, Santa Monica, Calif.: RAND Corporation, MR-1625-AF, 2002.

Galway, Lionel A., Robert S. Tripp, Timothy L. Ramey, and John G. Drew, *Supporting Expeditionary Aerospace Forces: New Agile Combat Support Postures*, Santa Monica, Calif.: RAND Corporation, MR-1075-AF, 2000.

Grunch, Col Russell, PACAF/LG-ALOC, electronic mail communication September 11, 2003.

Hanser, Lawrence M., Maren Leed, and Charles Robert Roll, *The Warfighting Capacity of Air Combat Command's Numbered Air Forces*, Santa Monica, Calif.: RAND Corporation, DB-297-AF, 2000.

Headquarters, Air Force Materiel Command, Virtual Inventory Control Point concept of operations, draft, February 6, 2003.

Headquarters, USAF/CVAE staff, transcript of interview, January 4, 2000.

Headquarters, USAFE/LGT staff, transcript of interview, February 1, 2000.

Joint Chiefs of Staff, Joint Publication 1-02, *Department of Defense Dictionary of Military and Associated Terms*, Department of Defense, April 12, 2001.

Kaplan, Robert D., "Supremacy by Stealth," *The Atlantic Monthly*, July/August 2003, pp. 65–83.

Koenig, Ed, et al., "Recommended Changes to Air Force Spares Programming, Budgeting, and Financial Processes," Programming and Financial Management Team, U.S. Air Force Spares Campaign, AF/ILS (Installations and Logistics), Washington, D.C., 2001.

Leftwich, James, Robert Tripp, Amanda Geller, Patrick Mills, Tom LaTourrette, and C. Robert Roll, Jr., *Supporting Expeditionary Aerospace Forces: An Operational Architecture for Combat Support Execution Planning and Control,* Santa Monica, Calif.: RAND Corporation, MR-1536-AF, 2002.

Peltz, Eric, Hyman L. Shulman, Robert S. Tripp, Timothy Ramey, Randy King, and John G. Drew, *Supporting Expeditionary Aerospace Forces: An Analysis of F-15 Avionics Options*, Santa Monica, Calif.: RAND Corporation, MR-1174-AF, 2000.

Rumsfeld, Donald H., *Defense Strategy Review*, June 21, 2001a.

———, *Guidance and Terms of Reference for the 2001 Quadrennial Defense Review*, June 22, 2001b.

———, *Quadrennial Defense Review Report*, Office of the Secretary of Defense, September 2001c.

Snyder, Don, and Patrick Mills, *A Methodology for Determining Air Force Deployment Requirements*, Santa Monica, Calif.: RAND Corporation, MG-176-AF, 2004.

Sullivan, Kevin, "Concept to Reality," *Air Force Journal of Logistics*, Vol. 27, No. 2, Summer 2003.

Tripp, Robert S., Lionel A. Galway, Paul S. Killingsworth, Eric Peltz, Timothy L. Ramey, and John G. Drew, *Supporting Expeditionary Aerospace Forces: An Integrated Strategic Agile Combat Support Planning Framework*, Santa Monica, Calif.: RAND Corporation, MR-1056-AF, 1999.

Tripp, Robert S., Kristin F. Lynch, John G. Drew, and Edward W. Chan, *Supporting Expeditionary Aerospace Forces: Lessons from Operation Enduring Freedom*, Santa Monica, Calif.: RAND Corporation, MR-1819-AF, 2004.

Zettler, Michael, "The New Vision," *Air Force Journal of Logistics*, Vol. 27, No. 2, Summer 2003.